WHAT THE BULLETS SAID:

Changing Rounds and Unstoppable Assailants

Richard Swiderski 2013

Dumdum Bullets

A working man, a little dumb,
Made for his boss a little gun,
A cartridge and a bullet
With point sawed off to dull it.

Another worker, just as dumb,
Made for another boss a gun,
A cartridge and a bullet
With point sawed off to dull it.

One day the poor dumb workers met,
Aimed at each other's wooden heads
And each one sent a bullet
With point sawed off to dull it.

Two bullets fled and said
"Dumb dumb,"
Two dummies tumbled over dead-
Never knew what the bullets said.

Oscar Ameringer (1914)

In the meane Tyme Salvages upon the Mayne did fall into discencyon wth Capte: MARTIN who seised the kings sonne and one other Indyand and broughte them bownde unto the Island where I was when a shipp Boye takeinge upp a Pistoll accidentyallie nott meaneinge any harme The pistoll suddenly fyered and shotte the Salvage prisoner into the Breste. And thereupon whatt wth his passyon and feare he broake the Cordes Asunder where wth he was tyed and did swimme over unto the mayne wth his wound bleedinge And there beinge great store of maize upon the Mayne I cowncelled Capteyne MARTIN to take possesyon thereof the wch he Refused pretendinge thatt he wolde nott putt his men into hassard and danger. So haveinge seene Capte: MARTIN well settled I Retourned wth Capte NELLSON to JAMES TOWNE ageine Acordinge to apoyntementte

George Percy, A Trewe Relacyon (1609-12: 266-67)

Table of Contents

Illustrations

Introduction

Each time a bullet is fired a shape is taken.

In June, 1676 a 12-year old London shop apprentice was convicted of manslaughter in a jury trial.[1] The boy had found a musket in his master's dining room upstairs from the shop. Adult men formed a civil guard in the city and the master had left the gun there after returning from the most recent muster. The boy pushed the end of a gunstick into the gun's muzzle and determined that there was a charge of gunpowder present. He scraped together a few grains of powder and put them into the firing pan. He then was called downstairs to the shop to attend to a customer.

Later returning upstairs, he carried the gun out onto the balcony and fired it into the street below. The gunstick remaining in the gun's barrel was propelled by the explosion into the chest of an elderly man walking in the street. He collapsed and soon expired. A group of people who had witnessed the shot came into the shop and apprehended the boy. His defense was that he did not know the gunstick was still in the barrel and he had no intention of harming anyone. However involuntarily he had killed the man, it was still manslaughter.

The apprentice's shot was the earliest case involving a gun recorded in the Old Bailey court annals (begun 1674). The record of the case hints that a heedless innocent firing a gun with deadly consequences was not unprecedented. The tragedy was made possible by the versatility of the gun, which fired the tool used to insert the projectile rather than the projectile itself. A surgeon serving as witness at the boy's trial confirmed that pieces of the gunstick were found lodged in the victim's body, which proved that it was a gunshot death and not due to any other cause. Children were imprisoned with adults during this period, but the record is silent about the final disposition of the apprentice's case.

There is a long history of children playfully aiming found guns at others, pulling the trigger anticipating only a blast and suffering the consequences. The equally accidental 1609 shooting at the English colony of Jamestown of a captive Indian man by a ship's boy was reported in exonerating language by George Percy.[2] Percy's relation also narrates the incredible strength of the wounded Indian breaking his bounds and swimming away, followed by the English raiding Indian maize supplies, which apparently did not ease the subsequent famine or prevent cannibalism.

In the Old Bailey case the boy only expected to disrupt the already noisy traffic in the street below. Perhaps he was even aiming the gun at the old man, out of the apprentice's inherent hostility toward

---

[1] Old Bailey Proceedings Online (www.oldbaileyonline.org, version 7.0. 30 December 2012) June, 1676, trial of boy (t16760628-4)
[2] Percy (1609-12: 266-67)

censorious elders. He proposed an explosion and shot out a missile. Gunpowder from its origins has been both recreational and projective, at separate times and at the same time. Gunpowder causes an explosion and is done whether or not a projectile is emitted.

This book is an examination of the projectiles made by gunpowder blasts, of what is unseen in the enveloping sound becoming seen in anticipation of the next shot. What does the shooter believe his shot will do, and how do those beliefs become embodied in the shape of the projectile? The enveloping percussion of the discharge induces certainty and inattentional blindness at the same time. Who is the shooter in terms of what he shoots: I wish to see that problem more clearly.

In the earliest firearms the explosion was the primary event, forcing out whatever was in the barrel. The explosion may have been the only event. Eventually the projectile was shaped both to pass through the tube and enter the target. Once primer, charge and projectile were packaged in a cartridge to form what we usually call a bullet (the term used for the projectile alone prior to the mid-19th century) the shape of the bullet became systematically tied to the expected result of the firing.

Concepts of the relationship between bullet shape and the outcome of the shot developed slowly in the direction of deliberate design to gain a certain outcome. It began with the managed explosion itself. A boy playing soldier in the late 17th century loaded powder and placed primer before pulling the trigger, causing the flint to spark and rendering the shocking noise. The projected gun tool was incidental. As all the components except the firing action itself became invested in the cartridge package, that entire package could be designed to satisfy immediate expectations. It could fulfill beliefs about what a projectile's shape and weight can do in a body.

In July, 2012 the Reuters news agency published online a video circulated by the Afghanistan Taliban, of the execution of Najiba, a young Afghan woman, by bullets fired from an AK-47.[3] The shooter is Jumma, her husband, who has already killed the young Taliban commander who abducted or eloped with Najiba. Jumma now fires a bullet into Najiba's back as she sits by the river draped in a shawl, and after that shot causes her to fall over backward, continues to fire one trigger pull after another. Men in the crowd standing about, some of them recording the event on their mobile phones, cry out, "Enough, enough!" then praise Jumma as a hero. Among these men are two other Taliban who took part in the gang rape of Najiba. Jumma approaches the lifeless body and continues to shoot; he obeys an enjoinder from the crowd to shoot her in the face. A reporter learned that Jumma now serves as the bodyguard of one of the rapists.

The bullets Jumma used were made for repeated rapid fire. The ability to fire them reliably and in quantity developed since the muskets of the 17th century. This allowed Jumma to confirm his separation from his wife and the family responsibilities she represented by freighting her body with more bullets than ever would be needed simply to kill her. The bullets have a fixed shape congenial to the mechanism of the gun but they change shape to become a mass pinning down his wife as the husband moves into the all-male society of her rapists. For the apprentice the explosion became a projectile; for Jumma the bullets became the departing mass of his wife. The two shooters authored separate blasts become solid and safely spectacular.

---

[3] Mogelson (2012)

The apprentice's shooting of a pedestrian is unlike the obliteration of the wife: a musket versus an automatic rifle; an unexpected projectile versus very much expected quantities of bullets; an accidental victim versus an intended one; an incarcerated boy versus an oddly liberated man; a single entertaining sound and many entertaining sounds. The wounded, fleeing Indian adds an accent to the bullet.

Bullets are cultural matter. Natural or manufactured they are shaped by the forces to which they are consigned. Bullets sent toward a target obey physical laws. From the point of view of any observer they obey other laws as well. How they are fashioned, aimed, impelled, guided and what they do upon being received into a body are described by chemistry and physics. How this is planned and understood is described by the culture and social allegiances of the fashioner, the victim and the viewer. Painstaking reconstructions of murders, fictional and genuine, are only one way values are added to the neutral physics of the projectile.

The point of fire and the target are connected with each other before the accident of the bullet.

Bullets are a good case study in the interface between the measureable physical effects of material forces and human perceptions cast by cultural categories. There are other physical interfaces with cultural patterns: weather and climate with human seasons, the flow of water with human settlement and travel, the movement of stars and planets with the passage of time, to name a few. Bullets are entirely a human creation made to mark an assault with a crescendo of sound.

Human design and intent cannot guide these forces entirely. The way bullets are perceived before, during and after use is an attempt to make up for this. From the structure of the gun, the materials and shape of the cartridge and bullet, the composition of the primer and powder, to the gunshot wound, the bullet is an attempt to channel matter and energy for human purposes. It is the prototype of sudden energy giving directed flight to a small projectile.

These purposes seem human. Given the nature of the bullet they are partisan human purposes. A bullet travels from one social location to another. The response to the bullet seems unambiguously a matter of flesh and blood. Yet it is a particular kind of flesh and blood, a pivot at the edge of physics and culture. A bullet taking shape will place its purpose in relief.

Bullets designed or believed to increase in size and change in shape upon contact with flesh, bring out three aspects of bullets. They are designed to attain a culturally specific target; they are used under a strictly defined set of cultural circumstances; and the results of their having been used are seen differently from one viewpoint to another. The physical makeup, flight and wounding pattern of the bullets is analyzed and remade in accordance with the beliefs of those concerned with them. Involuntary on the part of some parties at first, involvement with the bullets turns voluntary.

Bullets that change shape in the target are the subject of "motivated thinking," as psychologists term thinking driven by an agenda with diminishing reference to logic or scientific testing. Bullets as a technology invite culture to become a party to a conflict. Internal group coherence in opposition to other groups takes precedence over technology. Bullets designed to halt an attack by a group of people defined in opposition to the makers of the bullets can lead to an exchange between two parties with each party rationalizing its use of the bullets against the other. Shape changing bullets are aimed at someone who reciprocates. This exchange evolves into accusations hurled back and forth, and into a science of the evidence of their use.

Shape changing bullets when first conceived are aimed at another party considered deserving of their damage, then they are the source of wounds we identify in ourselves and finally they are an element in random injury and death, the violence of everyday life directed by the insane. The evidence of their making and use is at first subjective: the belief that the subject should send out a specific shape of bullet against a specific enemy. Then the evidence turns to the objective: bullets of a specific original shape have been sent out against by members of a hostile group. The written record progresses from defensive attack narratives to proposed proof of victimhood. From photographs of bullets to declarations of wounds. Each party can accuse the other of claiming to be the object when they really are the subject. They say they have been wounded by the bullets they make to fire against us.

As an object of production, and of mass production, bullets are the object that makes its own subject. Once the bullet is proposed as a means of destroying persons belonging to a certain category, bullets made in that design participate in the nature of the people against whom they have been made. The more they are produced the more the entire society of the producers is mobilized against the target people, who are ensnared by the reproduction. Individual fashioning of the bullets turns into an industry which leads to a market.

A bullet's cultural life progresses from a cultural stereotype guiding the bullet's initial shape to motivated thinking about what its final shape must be to the concealment of the motive in standard shaped bullets. The shape changing bullet appears to be a model for other pieces of technology made to permit the violence of their use to be justified according to social, economic and cultural divisions.

The shape changing of bullets is inherent in the nature of the gunpowder that propels them. Gunpowder's incendiary form is a component of projectiles sent to start fires, but as it grows more explosive it moves masses of small, damaging objects against masses of people then single weighty objects against individuals. The pattern of its introduction from surprise to receiving then sending is endlessly repeated. The receivers become the senders and all are enmeshed in the same network of explosions and projections.

Gunpowder invented among the Chinese was first used to complement and sharpen flame-throwing. The addition of projectiles to a burst created a spray that only later became confined to a tube and used to target objectives. The sound of the blast and the diffuse sparkle remained a potential within the use of gunpowder for projectiles that were forever being narrowed in focus and shape

The first introduction of firearms to people without firearms often came with invaders firing bullets at them followed by traders selling them guns and gunpowder. The provision of guns enabled the invaded to dominate and enslave others in their own vicinity, and they transferred the habits of existing projectile technology-slings and arrows-to gun use. They might give greater emphasis to targeting and marksmanship and less to the awe-inspiring effects of the gunpowder explosion. The American Indians, the Hawaiians and the Maori received guns from Europeans and turned them upon each other and upon the invaders. A systematics of exchange ensued with the issuing party trying to limit gun use and stay ahead of the recipients become emitters. The Japanese engaged in this exchange, stood back and inverted it in a wholehearted reprise.

An eighteenth century prototype of the machine gun fitted for both round and square bullets to be fired at Christians or Turks anticipates not the technology of shape-changing bullets but the unshakable position of self-defense in their use. Bullets of different shapes for different people do not change shape after they leave the barrel of the gun. They are expected to have their effect because they retain their shape. There never was a practical demonstration of this gun. It epitomizes thinking about bullets made for their targets that is mostly a suggestion for potential purchasers and investors in the arms of the nation, a precocious missile defense system.

Internal ballistics occupied the first bullet-makers. They were determined to get the projectile out of the gun tube at a maximum velocity. Early bullets were not much different from rocks or pieces of crockery. They sometimes were rocks or pieces of crockery. It became apparent that the tighter the seal between the bullet body and the wall of the gun tube the faster the bullet would be propelled out of the tube by the gases of the exploding charge.

An irregularly shaped piece did not guarantee a consistent expulsion even if safety was not a concern. Fitting a bullet's body to match the shape and diameter of a gun tube still left the need to load the bullet into the tube easily enough to permit exit and reloading. Musket balls were made spherical, but might be modified with roughening or notches to expand on contact with the target's surface.

Bullets were formed to expand in response to the heat and force of the charge, to expand enough to form a seal with the tube yet not so much that the tube became jammed. These were internally expanding bullets made of metal loaded into the muzzle, cleared out with firing and reloaded. Lead expands rapidly and consistently, and could be cast into shapes that allowed for a seal while facilitating travel along the barrel, but it had the disadvantage of melting into the surface of the barrel as it traveled, fouling the interior for further shots.

Contradictory interests were operating on the initial formation of balls muzzle loaded into guns. Rifling of the barrel might improve accuracy over the short range, but then the bullet would have to be shaped to spin consistently. Regularity in the shape of the projectile seemed to lead to improvement in performance not yet strictly measured.

Cartridges were at first holders for the powder poured into the gun barrel and a bullet that was rammed in afterward. Paper cartridges were torn open to yield one shot's worth of powder and primer and one bullet. Engineering a breech into the gun allowed bullet and powder to be loaded forward from the rear.

A solution to the problems of merely improved projectile design still not preventing fouling was to encase the bullet, its charge and the primer in a shell made of a metal with a higher melting point than the bullet metal. Brass often was chosen for the shell, lead for the bullet itself. The primer formerly was set off by a match or sparks from metal or flint and in turn ignited the charge loaded behind the bullet. Now the entire chain of firing was enclosed in the same package. The cartridge was held in the end of a gun barrel and shot out its lead bullet with the strike of a firing pin actuated by a trigger pull. The cartridge was then expelled from the gun and another cartridge was loaded.

Trigger and firing mechanics matured to maximize the advantages of the cartridge but that did not solve all the problems of internal ballistics. The cartridge itself did not travel the barrel but the bullet still did, and its design greatly mattered to how quickly it made its exit. The bore of the barrel, the ingredients and positioning of the powder, the metallic alloys that formed the casing and the bullet itself all were subject to experimentation and reinvention. The weight of the bullet and its force of travel determined the amount of energy it brought against any barrier it encountered. The nature of that barrier and what was behind it received and returned the energy to the bullet affecting further travel and shape.

Armies outfitted with guns required standardization. Operating principles derived from chemistry and physics were introduced to make it likely that many guns could fire at the same time, be reloaded and fire again.

The question of the relative efficacy of cartridge and bullet designs led to a consideration of external ballistics. The flight of the bullet and its range was a matter of the exit velocity, the explosive powder used, bullet weight and shape, gun structure, and the properties of projectile motion. Understanding the relationship between what happened to the bullet in the target and the design of the bullet itself demanded examination of shot targets, of bullet remains and experimentation with the perceived components of the shot. In the case of living bodies, which could not be well mimicked by non-living or dead materials, anatomy and physiology had to be called to witness. Treatises on gunshot wounds by physicians and anatomists appeared with the mass gun-bearing warfare of the late eighteenth century and multiplied as the wars of the following centuries yielded specimens for field surgeons.

Shooting at large, potentially dangerous animals and fleet animals that might escape after being shot to die elsewhere made hunters seek bullet designs most able to stop the animal with the first shot. How a bullet acted against an animal's skin and in an animal's interior determined its effectiveness. Bullets that penetrated the interior and then deformed to tear blood vessels or vital organs were the most effective. The size of the animal, its temper, hide and anatomy indicated a bullet type best able to subdue the animal and preserve the hunter. If the nose of the bullet was not covered by the casing it would deform sufficiently to stop large game.

This principle was not applied to human targets until cultural considerations intervened late in the century. The development of strong, smokeless powder giving greater velocity to the bullets coincided with a reduction in bullet and firearm size, to reduce the weight soldiers had to carry and make it possible to manufacture more from the same amount of material. The resultant increase in firing range increased the distance of engagement and the perimeter of defense. When these weapons were carried into battles against impassioned foes such as those encountered in colonial warfare of European nations, the soldiers came back with stories of firing many bullets into an attacker and still have him continue his assault. The difference in styles of combat came down to the firearm and bullet, which were considered defensive weapons.

Ghazis in Afghanistan, dervishes in the Sudan, Zulu in South Africa and fanatical fighters wherever the defenders of European sovereignty met opponents who not stop unless the first shot was decisive. Removing the metal jacket from the nose of small caliber bullets rendered them likely to deform into irregular bullets of larger caliber. The modified bullet was named after an arsenal and gunnery school near Calcutta, Dum Dum, where it was invented in the aftermath of a campaign in which the attackers did as much damage after not being stopped by small caliber bullets as they did with their own antiquated large caliber rifles.

The large caliber effect was thought to be necessary against people who did not know they should stop and be wounded when struck, a relic of an earlier gunnery that modernizing weaponeers wanted to destroy. American imperial troops encountering juramentados in the Philippines simply reverted to larger caliber bullets in the belief that they had greater stopping power than the new smaller calibers. They also tested dumdums in the field.

Dumdum bullets were mass produced from small rounds as cartridges with the metal jacket partially uncovered at the nose of the bullet. They were admittedly used in several campaigns by British colonial troops but then fell under a ban of all expanding and deforming bullets agreed upon by most of the attending nations at the 1899 Hague Peace Conference. Only Great Britain and the United States withheld their approval for that article, arguing that their troops needed protection from otherwise unstoppable marauders. Signatories to the agreement could not use "expanding" bullets against each other, or had to give a year's notice if they planned otherwise. The members of the conference were trying to preserve an etiquette in warfare they should have seen did not exist.

Tests showed, however, that the dumdum or soft nose bullet was not the only design that could cause gaping wounds. If the point of the bullet was hollow the arriving projectile also would deform and expand the wound cavity. A lighter metal such as aluminum incorporated into the point would cause the bullet to tumble on contact. The British Mark series bullets went from partially to fully jacketed hollow or light nose types. The light-weight pointed nose of the German spitzer bullet was known to promote tumbling as it crossed into a denser medium. As they did with the first dumdums soldiers might improvise by cutting a cross into the nose of the bullet.

In the early 20th century each European nation armed its troops with a bullet that technically was not a dumdum but might have an effect that was even more lethal. The dumdum category, initially including just improvised low caliber bullets with the jacket shaved at the nose, grew to include all bullets designed to expand upon impact. Questions were raised about the relationship between the subjective bullet shaped for a purpose and the objective bullet, known only from its wound, the state of the victim and possibly some recovered bullet fragments. Ideology trumped attempts to connect bullet shape with wound scientifically.

A pattern of charge and counter charge developed during the first 20th century war between a colonial nation and a sophisticated adversary, the sequence of actions fought by the British against the Boers, descendants of Dutch and Huguenot colonists, in South Africa. Both sides claimed to have captured supplies of dumdum bullets from the other and to have suffered wounds characteristic of dumdum bullets. The other sides were savages treating their enemies as savages. This only was exaggerated during World War I: the Allies and the Axis sought a propaganda edge at home and internationally with dumdum accusations while the bullets they were using might have the same effects.

Projectiles containing an explosive had been banned from "civilized warfare" by international agreement during the mid-19th century, yet combatants in wars tried to portray their enemies' ammunition as explosive. Bullets fragmented causing wounds prone to infection because of the number of lacerations. X-ray photographs dramatized the distribution of the fragments into evidence of blast-made shrapnel.

Externally expanding bullets were argued to be appropriate for firing into savages and others of a lower nervous organization, such as criminals and other members of the lower classes. The relative insensitivity to pain and emotion of the underclass made them less likely to be felled by projectile penetration. A civilized people could only assault another civilized people using humane weapons, which were meant to remove the fighter from the field of battle with minimum damage. Nations claimed their weapons were humane if they met this criterion.

The Japanese Empire entered upon the battleground of nations with a bullet its military called humane because it passed swiftly through the body of the adversary who faced only a brief recuperation. This was initially supported by some medical personnel in the field during the Russo-Japanese war. Evidence to the contrary was brought forward. Again it was a matter of applying a number of disparate parameters to ascertain the total effect of the bullet. How many killed? How many quickly disabled? The answers to these questions from mounting statistics vacated the idea of the humane bullet for most.

The changing shape of bullets was fixed in the character of wounds but could not be well predicted from the design and material of the bullet, nor could the nature of the bullet be determined from the wound itself. Yet verbal categories of dumdum, which contained a number of subcategories such as soft nose, hollow point and cross point, persisted in discourse about violent conflict.

The "treacherous Mexican soft nose bullets" discovered in the bodies of American victims were sold to the Mexicans by American and European firms, and smuggled over the border by profiteers during the Mexican war. From the first naming of dumdums during the 1890's the arms trade had provided everyone who wanted them with dumdums. The technique of making dumdums by scraping off the metal jacket or cutting a cross in the bullet point was readily learned and taught.

Expanding, deforming, or dumdum bullets of all kinds were available. The term dumdum, however, continued to mark an intention to do rapid and painful damage to stop the person struck. This template only became more intense during the 20th century.

A parallel development that anticipated the next stage in shape changing bullets was the alleged use of these bullets in assassinations. A political figure or a local leader was thought to be the victim of the sure kill of a dumdum bullet. Early in the 20th century the bullet-focused subjectivity of the expanding bullet's origins identified the assassin as someone who knew how to obtain or make the bullets. Later a sudden gunshot death even without an examination of wounds was sufficient to demonstrate dumdum use.

World War II drew upon memories of World War I dumdum accusations and beliefs. The charges and countercharges of World War I were echoed in the several actions leading up to the war but then seemed to disappear. Dumdums, their name echoing the throb of warfare, were a judgment and an explanation. Soldiers of a nation accused of using the illegal ammunition were executed due to that association. Episodes of alleged dumdum use surmised from wounds proved to be lethal for the prisoners of war from nations accused of the violation.

Winston Churchill had voiced the opinion of expansive bullets shared among the officer class in turn of the century colonial fighting. He accumulated the strains of the dumdum in his writings and opinions, finding them an appropriate stopping mechanism for savages and accusing enemy combatants of being savages for using them. On the brink of World War II, elected Prime Minister of Great Britain, he assumed the defensive pose tommy gun in his arms, and resisted comparison to American gangsters who in the period between the wars had become the savage dumdum users of the newspapers.

Dumdums explained deaths of individuals from severe bullet wounds administered by enemy snipers. Surgeons did not distinguish between battlefield wounds caused by jacketed ammunition deformed by ricocheting or causing severe wounds by tumbling at lower velocities and bullets that had been purposely altered to cause more severe wounds. They concentrated on the anatomical damage provoked by all bullets.

Since the early years of the 20th century expanding bullets had been undergoing an evolution that reshaped them more in accordance with the terminal ballistics being formulated. They were redesigned to expend their kinetic energy within the body of the target rather than carrying it on through and out. The design was to form a bullet that reliably stopped the advancing enemy. The bullets were formed according to an explicitly defensive intent.

The account of what the bullet actually does inside the body was coming closer to a verbal and visual representation of the moment of entry and travel. Bullet form retained the structures considered reliably expanding developed earlier in the century, but never successfully commercialized. Now these earlier modifications of the expanding bullet types were called upon to provide a visual image of the bullet's action in the body of the target. Bullets that deployed spinning blades of metal at front and that were not inhibited by the tendency of the hollow point to fill with intervening material offered fierce and dependable shape changes for the consumer personal defense market. The final phase of the round of ammunition is to turn the bullet's inescapability against ourselves. It might have been assumed that we were male, but faced with a dawning realization the bullets were used and made by women we tried to sharpen them for greater penetration.

The London apprentice pretending to be a soldier and the Afghan husband aspiring to the Taliban were separated from each other by centuries of cultural history, and united by the boundaries they affirmed with what emerged from their guns. The Indian no longer fled after being shot. It was an explosion taking projectile shape against the enemy who approaches.

1. The Give and Take

Chemical recipes included in a Chinese collection of military techniques (Northern Song Dynasty, 1044 CE) promise compounds that emit toxic fumes, send out barbs and start fires with a bang.[4] No one name is consistently used for these formulas, but the list of ingredients-always including sulfur, saltpeter and charcoal-reveals that they are forms of gunpowder. In none of them does the proportion of saltpeter exceed the 75% of total mass threshold necessary for the mixture to explode. They are incendiary compounds meant to surprise and disorient the enemy when delivered into their ranks.

Gunpowder before guns was a means of conveying an uncontrollable fire in warfare. Chinese alchemists had learned that they could "subdue" sulfur and saltpeter by igniting them and letting them burn together. Their caustic nature and inflammability then vanished in sputtering smoke that could disconcert assembled hosts receiving it on the tips of arrows. The addition of a small amount of charcoal sustained the burning and prevented premature ignition. It became apparent that the mixture burned even while enclosed in vessels or packages lobbed by catapults.

Chinese armies employed incendiary warfare from the earliest documented accounts. Fires were set and masses of burning material were delivered as projectiles or overland in carriages. Poisonous or explosive fires were started just by positioning the right selection of combustibles together. Among the animal delivery systems, birds carrying hollowed out nuts filled with burning material were released in the direction of the enemy.[5] Deflagrating gunpowder initially was an enhancement of the smoldering dried vegetable matter, the oil and naphtha used to carry fire to the enemy.

A 10th century painted silk banner brought to Paris among the cache of books and objects removed from the collection in the Dunhuang "Library" cave depicts the tempter Mara unleashing his demon hosts to distract a meditating Buddha. [6]

1. Detail from The Buddha Tempted by Mara. Hackin (1923: 35)

---

[4] Zhou (1986: 187-88); Needham (1986: 118-24) *Wu Jing Zhong Yao, Collection of the Most Important Military Techniques*
[5] Sawyer (2004:120-25)
[6] Hackin (1923: 35-39)

One of the demons, alert serpents emerging from his head, holds in his right hand a cylinder from which flames extend as the other hand shoves a rod into its core. Another serpent-inhabited demon is in the act of tossing a fiery sphere in the same direction, toward the seated Buddha. These warlike gestures are aimed at preventing the Buddha's enlightenment by drawing his attention to the noise and confusion of the world with its dancing animal heads and weapons.

The cylindrical flame thrower described in writings and seen in later representations is the *huo qiang*, or fire lance. In the temptation scene it seems only to be emitting flames, consistent with its history as an incendiary device pumping out naphtha. If the flames are from deflagrating gunpowder then it may also be spewing bits of metal and stone. Fired from an "eruptor," a pot or a bamboo tube, these objects were "co-viative":"because the gunpowder was not fully propellant, and the object or objects did not occlude the bore of the tube."[7] The objects traveled with the eruptive escape of gases rather than being directed at a single target. Force built up within the mixture because incendiary gunpowder burned within the confined space.

Gunpowder eruptors ejecting co-viative masses spread from China to the invading Mongols who carried them westward where they became a part of Arab, Turkish and European warfare. Gunpowder of each degree found uses. Burning gunpowder set fires. Deflagrating gunpowder was the impetus of fireworks and rockets. Exploding gunpowder shot out projectiles.

When saltpeter (KNO_3) burns with sulfur it releases oxygen that sustains the combustion. If the amount of saltpeter exceeds the 75% threshold the liberated oxygen feeds the flame to the point of a catastrophic release of hot gases, sulfur dioxide and nitrogen. The augmentation of the saltpeter portion in gunpowder in the mid-13th century made the mixture explosive, imparting greater kinetic energy to projectiles. The more uniform the mixture, the smaller the particles of each ingredient in contact with each other, the stronger the single burst of the explosion. Techniques for milling and consolidating gunpowder were devised to improve its force and transportability.

The principle of the gun developed in tubes fashioned to send out projectiles with the force of a single directed gunpowder blast. This was the form in which gunpowder was delivered by gunpowder possessors to those unacquainted with its effects. At first the burst and the sudden flame was frightening enough. To be convincing the blast had to do real damage, by setting fires and causing injuries. The gunpowder blast was realized in the gunshot that conveyed both shock and death.

Every human group the target of bullets overcomes shock and fear by acquiring the means of returning them with an enhanced certainty. The bullets are imaginatively revised to make them more likely to reach their target. Magic bullets, silver bullets, poisoned bullets appear and reappear in the annals of armed peoples. So do methods and means of warding them off and of ensuring that the aimed-for target always will be hit. The magic bullet, borrowed from folklore, has become a common metaphor for any quickly comprehensive solution to a nagging problem.

---

[7] Needham (1987: 296)

One story of the demise of King Charles XII of Sweden at the siege of the fortress of Fredriksten in 1718 was that a soldier in his own army shot him through the head with a button from the king's own coat.[8] Charles was reputed to be so strong that he had hardened himself against the force of projectiles: the "berries" of the fallen away bullets made his boots uncomfortable to walk in. Someone had welded together the two halves of the button to form a brass sphere with a lead center. Only a piece of personal property refashioned into a musket ball and shot with good aim could overcome the king's resistance and wound him fatally. Only a magic bullet that already belonged to him could enter the iron king.

Disadvantaged recipients of bullets overcome their disadvantages by making the bullets' intended operation ever more forceful in the other direction. The magic bullet is a grand reflection of the mundane course of bullets. This give and take reciprocates technologically and symbolically, shooter channeling the bullet to the specific target while at the same time treating the bullet as a commodity to be sold to the target who buys it to send it back. The compound nature of gunpowder-the surprise, the fire, the penetrating explosion-activates this reversible exchange. The size of the projectile increases and diminishes. The balance point of this exchange is reached in the single bullet fired by one individual member of a group against another individual member of a group. Bullets are aimed and returned across social and cultural boundaries.

The delivery of gunpowder and metal products to those who did not make them was one of the thrusts of the colonial advance from the beginning of the 16th century. It was a military and a trade initiative. The firearms-making nations exercised that seeming advantage over others who had never witnessed explosive projectile weapons. The result of this incursion was the spread of firearms through trade and capture to the rest of the world. The scene of attack, defense and reception of firearms technology was repeated in the Americas, Africa, Australia, and the Pacific beginning a process that repeated itself within those continents as one people took up firearms and introduced others to their use.

Those with imperial ambitions, the desire to arrogate the resources of lands with susceptible populations, used firearms as offensive instruments and to seduce existing polities into the cycle of domination through weaponry. The movement of people this initiated contributed to the spread of diseases and systems of enslavement. Populations on the west coast of Africa, on the east coast of America and in Madagascar were among those who engaged in the arms exchange with European interlopers.

American Indians saw the advantages of bullets for hunting: the projectiles were not deflected by intervening vegetation as arrows might be. The potential loss of surprise during the loading and firing was compensated for by the greater likelihood of a kill. Guns for hunting also were guns for warfare. Guns were used to recruit Indians into waging war on behalf of the gun providers. Colonial administrators tried to regulate Indian acquisition of firearms.[9] Sellers invented ornate lower quality guns for the Indian trade. The Indians could sharpen their own flints and shape their own musket balls but they remained dependent upon European and American manufacturing for their guns and gunpowder. Cultural and social differences between Indian groups now included gun acquisition and skills.

---

[8] Dash (2012)
[9] Starkey (1998: 20-22)

As the colonists occupied and defended more land, the European tactic of massed fire from a concentrated block of gunmen cumulatively proved more lethal than the Indian stealth attack. The rifled bore of the defenders' firearms more than compensated for the Indians' better marksmanship with their relatively inaccurate smooth bore pieces.[10]

The battlefield at Little Big Horn, the last major armed confrontation between Indians and United States government forces, contained bullets fired by the Indians from a variety of guns, many of them antiquated.[11] The U.S. cavalry fired mostly Model 1873 Colt .45 revolvers and Model 1873 Springfield .45 carbines. The Indians had not succeeded in arming themselves exclusively with the most advanced weapons of the time. They did prevail on that battleground, but not afterward.

The introduction of gunpowder and guns to those who did not have any such weapon initiated a give and take in which the sides were never fully reversed. The novices received gunpowder and guns but never were able to originate either. One precociously armed group might defeat others in their vicinity to create a kingdom, but that kingdom fell to the gun suppliers and was incorporated into a larger international empire. Representatives of the empire were always at risk of being put in the defensive role against better armed native subjects. The first shock of gunpowder had to be channeled into gunpowder projectiles aimed at those with projectiles of their own.

The experience that was enshrined for many people newly encountering guns delivered both by capture and in trade with colonial invaders was somewhere between that of the American Indians and that of the Maori. Muskets brought to New Zealand by colonists from England were adopted by Maori for their internecine struggles. In the earliest recorded confrontation between a party of Maori armed with muskets and traditionally armed Maori opponents the musket bearers were defeated when they paused to reload.[12]

This was only a temporary setback for the progress of gunpowder weapons among the Maori. The "Musket Wars" on 1818-36 followed with more lives taken by gunfire than had been possible with stone and wooden weapons. By 1845 a colonial administrator promulgated an Arms Ordinance to curtail the adoption of firearms by the warring groups.[13] But that would have curtailed sales.

Both the race by the arms originators to stay ahead of the colonized in technical prowess and the unpredictable encounter between the gun-armed and the sword- or cudgel- armed were taken up in the current of relations between colonizers and colonized. Magic was one approach to try to gain an advantage but magic always reaching into the reality of bullets to try to improve their construction.

The Western Apache applied a caustic mixture of ground nettles, lichen and rotten deer spleen to their bullets.[14] Some Maori used paper Bible pages to patch their bullets and inscribed Biblical passages on the stocks of their guns.[15] The Jesuit missionary Ferdinand Verbiest wrote the name of a saint and that of Jesus in Roman letters on the barrel of each cannon he cast for the Qing emperor. James Puckle and Kyle Tunis

---

[10] Chet (2003: 56-59)
[11] Scott, et al. (1989: 105)
[12] Knight (2009: 24). The battle of Moremonoi in 1808.
[13] Mitchell and Mitchell (2007: 406-07)
[14] Jones (2007: 49)
[15] Haami (2004: 19)

offered spherical bullets for the Christian enemy and cubical bullets for the Turkish enemy. The intent and structure of physical modification of the bullets varied from one setting to another. They all were aimed at striking the enemy more strongly and more certainly than by the force of the bullet alone.

Before there was gunpowder the ancient Greeks, the Romans and their enemies flung stone, ceramic and metal projectiles at each other with slings.[16] The sling bullets were a made in various rounded shapes, some anticipating ball-shaped, ogival and rounded nose bullets fired from guns. These sling bullets were cast from stone or ceramic molds, which have been recovered in archaeological excavations. A few bullets were inscribed with names, exhortations to speed, taunts (some obscene) and obscure lettering.

These customs of enhancing projectiles, stone points and metal slugs, through language and medicine to influence their physical performance were passed down to bullets and to the pieces that fired them. The adversaries facing each other armed primarily with guns and other gunpowder weapons may have looked across cultural and social divides but they prepared what they sent against the other in the same way. The Turks fired bullets at the Christians as the Christians fired bullets at them and at each other. The Christians needed an edge over Turkish bullets acquired from the same sources.

Mass production of bullets, whether to be flung from slings or shot from guns, stamped the bullets with a uniformity specialized for the group making them. This allowed the bullets to be sold to the groups they were at first aimed against. National bullets were modified into articles of commerce. In the late 19th century ordnance experts formulated lists of slightly varying bullet types by nation: shorter or longer casings, differing noses, presence or absence of cannelures. During the World War I these same nations accused each other of stockpiling dumdum bullets that could be made from any of these types. After the war the Italians accused the British of selling dumdum bullets to the Ethiopians.

Captain James Cook shot at the Hawaiians pursuing him across the beach as he tried to reach a boat to return to his ship. Before he could fire another blast from his two-barreled musket the warriors cut him down. Kamehameha accumulated arms through trade and brought most of the Hawaiian Islands under his control through warfare and intimidation, spreading guns and bullets as he went. The kingdom Kamehameha created became a United States overseas territory on the same side of the firing line as the United States when the Japanese attacked Pearl Harbor.

The enemy who learns of bullets by being their target often buys them in quantity from the one doing the firing. That initiates competitive cooperation between the parties.

This pattern of defensive firing of bullets against an adversary who then became the firer's customer built a pressure for the group shooting the bullets to stay ahead of the recipient. Preventing the recipient from buying bullets would be contrary to the commercial interests of the emitter who also was a seller. The receiver of the bullets might use them against the first shooter or others and reinitialize the pattern. That in turn put pressure on the first shooter to innovate again, to get the better of his customer and offer a new product.

The Japanese received firing tubes and gunpowder from China, possibly during the attempted Mongol invasion in 1281, but they did not adopt firearms as weaponry until a daimyo purchased matchlocks from

---

[16] Pritchett (1991: 39-53)

Portuguese sailors on a Chinese ship driven by a storm to the southern island of Tanegashima in 1543. A chronicle of the events written 60 years later recalled the impression the firing of the gun by the Portuguese made on the Japanese onlookers. It could "make a mountain of silver crumble and break through a wall of iron. Someone with aggression in mind toward a neighboring country would lose his life instantly when hit." [17] The same for deer caught eating rice from Japanese plantings. The overwhelming force of bullets, their seeming irresistible accuracy, impressed itself upon the rulers and the ruled in Japan.

Artisans made copies of the guns and the warrior class became expert in firing them. Soon they proved their prowess in helping the Tanegashima lord overcome rebellious subjects and in the course causing the first firearm deaths in Japan. Within twenty years battles between rival lords were being fought with firearms. Japanese forces that invaded Korea (1592-98) battled Chinese troops also armed with muskets and cannons.

An attempt to get metal weapons, swords and guns, out of the hands of peasants and merchants, accompanied granting the warrior caste, the samurai, a monopoly on armed violence as long as they were constrained by loyalty to the central government of the shoguns. The founding shogun, Hideyoshi, called upon the peasants to contribute their metal weapons to be forged into the image of the Buddha. The Japanese did not give up the gun, or even put it aside. The sword remained the primary weapon until the beginning of the 19th century saw the arrival of European ships with gunnery the existing powers could not ignore.

When fired upon the Japanese purchased new designs of gunnery from the interlopers. The armed forces of the shogunate, overwhelmed by a movement to restore the emperor, still fired heritage matchlocks of the Tanegashima design among the other archaic guns in their armament. The Japanese were so rapidly enrolled in the competitive exchange of bullet designs that their bullets figured among the European models in profile illustrations of national bullets that accompanied ordnance treatises in the late 19th century. Having passed through a long phase something like American Indians' acquisition of arms and bullets, they merged with the other armed industrial powers in the exchange.

The Japanese were enrolled in the national scheme of bulletry later than the Turks, who had developed bullets of their own out of the same international traditions as the Europeans.

Minié balls were made to maximize the force of expanding gases pushing the bullet forward through the gun barrel. Cartridges packaged the gunpowder to save time in the act of loading, and jacketing was to prevent the more rapidly forward moving lead mass from making direct contact with the bore of the gun. Within these mechanical innovations there was latitude for slight revisions thought to improve integrity and thrust. Toward the end of the century there was an open field for idiosyncratic variations that did not impinge seriously on the achieved efficient profile of the bullet, but were not just ornaments.

In the conflicts between armed nations an impetus was arising to make the exchange of bullets an end game simultaneously military, political and technological. Shaping the bullet against a foe was the beginning of the continuing end. As one party sought a technical edge that would make the bullets into the deciding factor they introduced elements into bullet design that would seem to make them especially effective against this particular foe. The bullets were entered into the give and take, the sale and purchase, of all projectiles

[17] Lidin (2002: 38)

with a point crafted for some particular other. The rules projected an efficacy that never was verified. Gunpowder begun with surprise and flames proceeded into a projectile that carried them both into the foreign body.

The shaping and reshaping of the bullets themselves is the lead point incorporating all the elements of gunpowder in an alchemical mix of matter and spirit. The cultural and economic surroundings of this pattern of introductions and productions encompass an originator who engages a group of recipients who return fire. This pattern is an action and a narrative framework making gun introductions to seemingly naïve recipients seem intelligible. The anthropologist Napoleon Chagnon became enmeshed in a controversy about his fieldwork among the Yanomamo in Venezuela that included accusations of using and supplying shotguns to heighten intergroup violence.[18] The charges and countercharges that spread over years seemed to place Chagnon and other parties in the same relationship with the Yanomamo as earlier suppliers of guns and ammunitions held to those without a previous source. Bullets, or in this commerce shotgun shells, took shape as they were exchanged.

---

[18] Chagnon (2013) accuses Salesian missionaries in the late 1960's of providing individual Yanamamo with shotguns they used to kill adversaries in raids. He himself had given the Yanomamo guns and ammunition, but not for raiding.

2. Turks and Christians

A difference between the "Christian" body and the "Turkish" body, at least when they met in combat, was that Turks could be stopped by square bullets, whereas rounds were sufficient for Christians. This expectation was advanced in an English patent (A.D. 1718 May 15 No. 418) for "A Portable Gun or Machine called a Defence, that discharges so often and so many bullets as renders it next to impossible to carry any ship by boarding." On the patent sheet a broader range of defense possibilities is listed: "for Bridges, Breaches, Lines and Passes, Ships, Boats, Houses and Other Places."

The patent diagram for this tripod-mounted ancestor (in sentiment) of the revolver and machine gun includes a detachable hand-cranked round bullet plate to deter the Christians and a separate square bullet plate aimed at the Turks. Such equanimity in projectile screening is rarely encountered in the history of firearms.

The precise mechanism was not delineated in the patent application.[19] Accounts of the gun's operation come from the key to the numbered components that accompanied the patent (below), and from analyzing surviving models of the gun held in various museums.[20]

Powder and bullets (14 and 20, spherical and cubical) were loaded into the corresponding cartridge (18 and 19), the cartridges were mounted each in a chamber in a circular plate with openings shaped square or round (16 and 17). The plate-cartridge assembly was attached on a spindle (3) to a base plate (15) on which was threaded a hand crank to the rear. The crank turned the assembly of loaded chambers to bring a cartridge into the breach (2). There the cartridge locked to hold in gases as a lever mechanism struck a flint producing sparks that set off the powder of each cartridge.

The plates in the diagram have six chambers but a photograph of a model shows nine chambers on the plate. The patent diagram includes an image of a bullet gang-mold (21) and a charge block of 20 cubical bullets (13).

The firing was sequential rather than a simultaneous volley: the crank was turned and the bullet was fired. Loading and firing were separate operations. The gun with its single barrel and lateral range of motion was an improvement over the weighty many-barreled volley firing flintlock revolvers that already existed at the time.[21]

John Ellis comments in his social history of the machine gun that the inventor James Puckle (1667-1724) made no provision for a change of barrel between square and round to accommodate the two bullet shapes, which places Puckle less in the history of machine gun technology and more in the history of machine gun promotion.[22]

---

[19] Dixie (1921) "some obscurity regarding its actual mechanism" Includes the patent drawing and a transcription of the text.
[20] For example the model at Boughton House Museum:
www.boughtonhouse.org.uk/htm/tour/revolverinfor.htm
[21] Willibanks (2004: 22-23)
[22] Ellis (1975: 13-14)

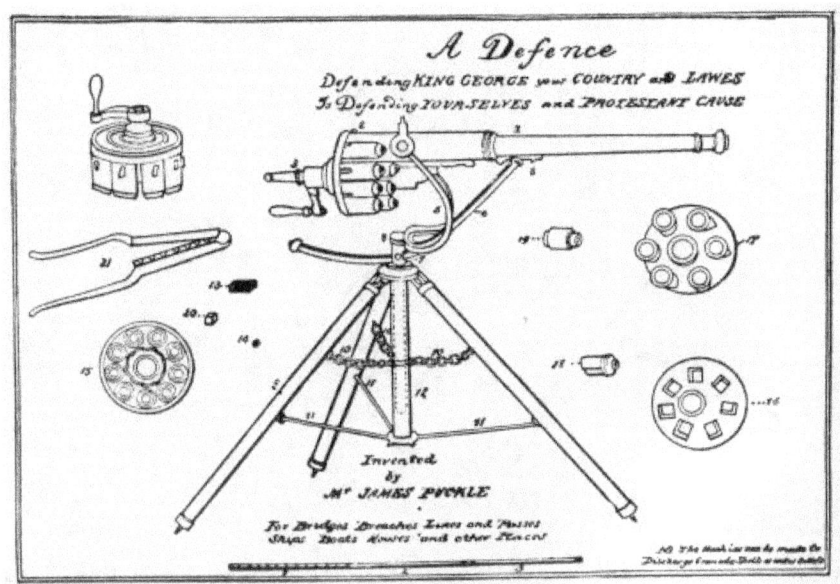

2. *Puckle's Defence patent drawing. Dixie (1921)*

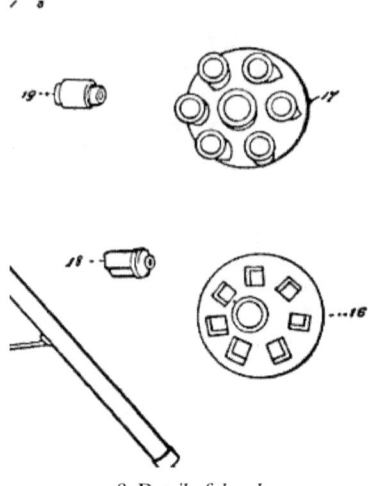

3. Detail of the above.

Puckle was a London solicitor who geared his patent to the new official requirement that the actual workings of the proposed device be described for the patent to be entered. He earlier had occupied himself in addressing the economic development of his island nation with tracts on woolen manufactures and fisheries. In a published dialogue between an Englishman and a Dutchman he proposed expansion of the English fishing fleet as a means of increasing Britain's economic and social security.[23] Puckle was a Projector, a proposer of schemes meant to lure investors. The year Puckle opened his Defence project for shares (paid in at 4 pounds sterling and sold at 8 pounds) also was the year of the South Sea Bubble investment fiasco. A verse epigram satirizing Puckle's project in *The Bubbler's Mirrour: Or England's Folly* (1720) concluded: "They're only Wounded that have Shares Therein."[24]

The Defence came late in Puckle's career, which had included projecting another machine, for making linen cloth. His biography was written, not because of his gun, but in the aftermath of the discovery that he was the author of an anonymously published collection of character sketches, *The Club.*

The gun was tested, and according to the *London Journal* of March 31, 1722 achieved the rate of 63 firings of one large ball or 16 musket balls in 7 minutes, in the rain with a flintlock action no less.[25] A few of the guns were assembled and linger as originals or replicas in house museums and in the Tower of London armory collection. There is no record of their having been adopted by the navy and aimed and fired at Christians or Turks boarding a ship or under any other circumstances. There is no record of any of the existing models being test-fired, but there is at present a video on YouTube of a miniature replica being fired by touching the primer with a soldering iron match.[26]

---

[23] Puckle (1718)
[24] Stephens (1873: 429). Quoted by Dobson (1896: 273), the best source on Puckle biography.
[25] Ffoulkes and Cottesloe (2011: 34)
[26] www.youtube.com/watch?v=8nTqV7o2jE8

The gun's capabilities, sorely needed in the 18th century battle of nations, were wishful thinking. The existing flintlocks took so much time to load and reload that only as concentrated defensive fire were they effective against onrushing crowds armed with faster weapons like spears and swords. Puckle's Defence promised to meet the attackers with unhesitating fire operated by at most two soldiers from a single barrel on a swivel base. A rapid firing mechanism would be reassuring to defenders faced with furious boarding parties. "The plate of the Chambers of the Gun for a ship shooting Square Bullet against Turks," reads the patent key note for number 16. Puckle incorporated an imagery of the gun's success into the technical description itself.

After the initial promotion, the Defence fell into obscurity for a hundred years, only to be revived as an example of the foolery being sold at the time of the South Sea Bubble. Charles Mackay, quoting the derogatory *Bubbler's Mirrour* epigram and citing the gun's inclusion on a set of satirical playing cards, grouped it with early 18th century financial follies.[27] P.T. Barnum listed "Puckle's Machine Company" among the "the humbugs of the world" for its gun "for discharging cannon-balls both round and square." [28] These two historians of gullibility led a procession of writers on deluded investment that continues past the early 21st century financial meltdowns to the present day.

Yet Puckle was working in the midst of attempts to design and manufacture rapid fire gunnery whatever the shape of the bullets. In his time colonial powers already were beginning to reflect on their own defensive capabilities against the firearms they were deliberately and accidentally introducing among the colonized. The shape of the bullets and the rapidity of fire were both points to be considered.

A review of 262 firearms patents filed in England between 1718 and 1852 found 9 proposals for "revolving or rotating firearms" among the many other single shot pieces.[29] Stimulated by Puckle, the effort to machine rapid fire continued but did not take a practical form until Gatling's 200 rounds per minute device (1862).

The co-inventor of the square bullets was the emblematic-sounding Kyle Tunis. Puckle's rationale for the impractical versatility of the weapon was that a volley of (square) bullets would convince the Turks of "the benefits of civilization." Presumably this meant that anyone who could conceive such a defensive weapon was civilized in a manner to be emulated. And according to Puckle the Turks, who had their own gun and bullet-making industries, were not civilized. There is no other record of square bullets being designed and produced. The inventors must have thought they would do more damage than round bullets, which makes Puckle and Tunis heralds of shape-changing projectiles' stopping power when shot into a raging body on the attack.

---

[27] The eight of spades. Mackay (1856: 1,61). An examination of the South Sea Bubble card deck viewable online at http://www.library.hbs.edu/hc/ssb/recreationandarts/cards does not turn up the Puckle reference, but Mackay and others may have been looking at another deck.
[28] Barnum (1866: 217)
[29] Firearms and Projectiles: Recorded Patents, *Journal of the Society of Arts* 2(1854): 601-2.

The difference between square and round bullets could easily be conceived "a delicate distinction which I am afraid would hardly be appreciated by the recipients."[30] How might the bullet shape difference strike potential investors?

From a close examination of the patent drawing the square sided, flat nose bullets appear to be mounted in a short cylindrical base, not exactly a cartridge. If they had a conception of bullet spin Puckle and Kunis may have thought of the square projectile shearing into the Turk's armor. Its flat head, unlike the spherical balls that were the projectiles of the day, was to slam into that attacking body with a decisive blow. The flat headed rounded bullet Puckle reserved for the Christians eventually became standard service bullet and remains an option for handguns and rifles. Of all the many shapes bullets have taken cubical bullets appeared only in Puckle's patent.[31]

The commentator in the 1854 review of gun patents set down the judgment: "The invention of James Puckle is that of a humorist..." This meant of course that the invention was amusing and curious; it also meant that it was devised for the humoral body. That body was composed of four fluids-blood, phlegm, black and yellow bile-each a combination of hot-cold with wet-dry, in quantities and locations that set the character and health of the person. Each humor was influenced momentarily and in the long term by outside influences: star and planet alignments, weather, landscape, food consumed and, no doubt, the shape of projectiles entering it.

The generally choleric (hot, dry) Christians could be halted by a round lead; the phlegmatic (cold, wet) Turks would need a square one. "When the customary sedateness of his temper is ruffled," an encyclopedia entry on the phlegmatic Turks reads, "he seems possessed with all the ungovernable fury of a multitude..."[32] This body lingered in medicine, politics and popular expression long after a more scientific, less cosmological anatomy was adopted. It set a persistent pattern reflected in the initial shape of the bullet addressed to the collective temperament of the target.

The threat of that body as Turkish was increasingly remote for an Englishman of Puckle's generation. The Ottoman Empire, having failed in a second attempt to take Vienna in 1683, was forced to cede European territorial possessions to rising European powers and adopt a defensive posture, especially against the ambitions of the Russian Empire. The Turkish threat was a sales pitch, a stand-in for any unpredictable foreign bodies who might throw themselves at Christian defenders.

The one documented plan to deploy Puckle guns was part of a colonial enterprise. In 1722 King George I granted John, the Second Duke of Montagu a royal charter to settle the islands of St. Lucia and St. Vincent in the Caribbean (the West Indies).[33] Montagu set up a company and recruited settlers with the prospect of planting sugar cane, indigo and other tropical crops of value as exports. He constructed a port and shipbuilding center at Bucklers Hard on the Beaulieu River in Hampshire, near a forest that would serve as a source of shipbuilding materials. Two Puckle guns were ordered from the Puckle Machine Company.

---

[30] Latham (1866: 95)
[31] Lash (1958), without mentioning Puckle, treats the square bullet as a useless invention accepted by the U.S.Patent Office in its earlier years.
[32] Brewster, ed. (1832: 98)
[33] Taylor (2012: 43-45)

The four ships carrying the first Montagu settlers to the islands were unable to establish a beachhead among the Indians, Africans and French entrepreneurs already present. The captain of one of the ships, Nathaniel Uring, recounted the inhabitants' firm resistance in the face of Montagu's proclamation.[34] Firearms were displayed and demonstratively fired without injury. Neither Uring's nor any other account of the expedition mentions the Puckle guns.

The Duke lost a fortune. His descendants developed the shipbuilding enterprise and later a yacht basin at the port. Montagu's two Puckle guns are on exhibit, one at a maritime heritage museum in the former ducal palace at Beaulieu and the other at one of the Duke's great houses.[35] The museum exhibit places the gun in the diorama of a sandy beach lined with palm trees. The second Duke of Montagu was a notable practical joker but the Puckle gun on the beach was no joke.

The guns themselves may never have been used but their plan and their name precipitated an intention. The aggressors thought of themselves as defenders, of the Crown and Nation, of their property acquired abroad and their own lives.

Those foreign bodies were advancing upon European and American colonialists in ever greater numbers as the 19th century progressed, and the existing means of stopping them seemed feeble. Of course the colonialists were not warding off invaders. They were resisting the attacks upon their outposts by Zulus, Somalis, followers of the Mahdi, Indian hill tribes, Maori, Sioux and Comanches to name a few-not always resisted successfully despite the invading defenders' superior weapons. Puckle embodied in his weapon the patriotic defense of ground wherever that ground may be, the need for overwhelming rapid fire and the fantasy of greater stopping power engineered into projectile shape shot against those hurtling others.

---

[34] Uring (1726)
[35] The Bucklers Hard Maritime Museum, Beaulieu, Hampshire. The Duke of Montagu also was the owner of Boughton House, now a museum where another Puckle gun is on exhibit.

3. Gunshot Wounds

What a projectile was seen to do to the body it struck derived from beliefs about the projectile and about the body. The appearance of the gunshot wound as a distinct type of injury coincided with increased focus on the shape of a projectile designed to be forcibly emitted from its source to enter an imagined body. This body anticipated the projectile's contours. The development of treatment for gunshot wounds discloses the body taking shape around the projectile.

Europe's internecine struggles of the 15th through the 17th centuries put pressure on rulers to field larger armies more readily gathered and trained, and to rely less on highly skilled archers and armored fighters and more upon expendable peasant musters. A weapon that would increase the force of less expensive troops was always welcome. Gunpowder and firearms had originated in China, and had entered Europe in the late 14th century in the form of metal tubes that held a projectile and a charge of gunpowder touched off through a hole. The first written instructions on the treatment of gunshot wounds coincide with the introduction of matchlock trigger mechanisms. The growing attention devoted to gunshot wounds from Pfolsfeundt in 1460 to Paré 100 years later reflects the increasing use of matchlock weapons and the demystification of gunpowder.

European battlefield surgeons began to distinguish treatment for gunpowder projectile wounds from treatment for other wounds based on their experience during the wars of the late 15th century, when gunpowder- impelled projectiles joined the other hazards of warfare. Heinrich von Pfolsfeundt in a 1460 surgery manual written in German (rediscovered and printed in 1868) advised locating the embedded projectile with an iron probe.[36] Pfolsfeundt prefaced his manual with admonitions on the personal hygiene, moral state (go to Mass) and deportment of the surgeon. Like his predecessors and followers he recommended narcotic inhalations and "wound drinks" to calm and strengthen the patients for all procedures.

A late 15th century document sets out recipes for a wound drink and a plaster specifically for gunshot victims.[37] Savin juniper, bulrush, mugwort, and dittany are the herbal ingredients of the drink, demulcent and diuretic, meant to promote urination and curb inflammation. The poisonous savin juniper, a common ornamental landscaping plant today, added force to the purge accomplished. As a wound drink the concoction served to carry away what had entered the body as violently as it had entered. That was chief purpose of projectile treatment: the projectile was a foreign body forcibly injected and forcibly removed.

The instrument of projectile firing was the bombard, a stationary cannon, a thick metal tube loaded with gunpowder and shot at the muzzle and touched off with a piece of burning cord (match) at a hole in the rear. Matchlock guns were hand-held versions of the same device. The blast, if it didn't detonate backwards, expelled the projectile and a mass of unignited gunpowder. The particles entered the wound together with the projectile itself. As breech (rear opening) loading systems were developed for guns the dangers of implosion and the expulsion of gunpowder continued to bedevil soldiers and surgeons.

---

[36] Garrison (1922: 193n) original German text.
[37] Vollmuth and Kronabel (2011)

The gunshot wound section of Hieronymus Brunschwig's 1497 surgical manual calls for a strand of fiber, a seton, to be drawn through the wound to clear it of poisonous particles and stimulate suppuration, to be followed by enlargement of the opening and removal of the foreign matter with forceps. The surgeon must be able to distinguish an arrow wound from a gunpowder projectile wound, know whether it is metal or stone that is embedded, whether he should cut it out, draw it out with poultices or soften the spot to permit it to come out on its own.[38] The object was to restore the body to its proper arrangement of parts after the disturbance.

Brunschwig's *Buch* was compiled from classical, medieval, and Arabic texts, and the author's own experience, which must have been the source for the gunshot wound treatment. It was translated from the original High German into Latin and several other European languages, including English and Czech. It contains a reminder that metal was only one of the materials shot from early firearms. Any object that could withstand the initial blast and travel a trajectory to penetrate a target could be loaded into a firing tube. Knowing what was in the wound was critical to the surgeon's success.

These early manuals were elements in a growing compendium of advice shared among surgeons who encountered many kinds of wound. The increase in guns and gunshot wounds gave surgeons more opportunities to test wound treatments and communicate practices if not results.

The practice recorded by the papal surgeon Giovanni da Vigo in his 1514 *Copiosa* also depended upon the belief that a gunshot wound was a case of poisoning.[39] The poisoning and the resultant decrepitude of the wound was caused by the gunpowder entering the flesh, and especially by the saltpeter in the gunpowder, known to be corrosive.

> If the wound was caused by the instrument called the bombard, cauterization should be done with the oil of elder or linseed (seminelinus) oil in the same manner. For the next three days the wound is covered with Egyptian unguent, prepared according to our recipe without arsenic. Over all the limbs according to the danger of cankers for several days the following plaster is spread...

da Vigo's treatment is often portrayed in the medical history literature as a bath of boiling oil and little else: a lethally traumatic treatment. The oil was to close the blood vessels communicating with the wound and to neutralize the saltpeter. The Egyptian unguent, the formula for which da Vigo gives elsewhere in his writings, was a treacle, a honey and vinegar softening agent intended to draw out the foreign matter. The flour-based plaster completed the extractive action of the surgeon's applications.

da Vigo's Latin text was translated into European vernaculars (Italian, French, Spanish, English) during the following century and corresponds to Brunschwig's earlier manual in the treatment of battlefield firearm injuries, reflecting a general consensus about the approach to these wounds. A difference between da Vigo, Brunschwig and Hans von Gersdorff, another military surgeon who retired from battlefield medicine into private practice, is that von Gersdorff presents his treatments as the result of actual field experience. His book titled *Feldtbuch der Wundartzney, Field Manual of Wound Doctoring* (1517) does not reproduce the orientation of the previous manuals and compile past texts on one wound or another. von Gersdorff does

---

[38] Brunschwig's *Buch der Wund-Arznei* section translated in Zimmerman and Veith (1993: 209)
[39] da Vigo (1531: Lib. primus de vulneribus, ca. viii de vulnere venenoso)

not go so far as to recount specific cases, but only enumerates the number of operations he performed on specific injuries with circumstantial details on variations actually encountered.

Previous writers on gunshot wounds had only their own experience and that of contemporaries to consult. None of the classical writers on surgery addressed gunshot wounds. Their strictures on projectile wounds were adapted, and the medicinal preparations such as wound drinks and plasters applied to any wound. Immediately after his section on wounds caused by the bombard da Vigo offers similar advice on arrow injuries. The second book of von Gersdorff's *Feldtbuch* is prefaced by an engraving titled *Wundenmann (Woundman)* : a standing male is struck, stabbed or shot about his body by cudgels, truncheons, bludgeons, knives, daggers, swords and arrows. von Gersdorff's gunshot treatment is in this section but an exploding tube is not among the instruments attacking the man.[40]

von Gersdorff breaks from the established treatment in not emphasizing that the wound has been poisoned by gunpowder. Like Brunschwig, he widens the wound channel. Instead of hot oil, he infuses the site with warm linseed oil and extracts the bullet using a variety of metal points. He learned this skill from Master Nicholaus, known as the Mulartzt, the battlefield surgeon attached to the forces of Duke Sigismund of Austria during the 1476-77 Burgundian Wars.[41] He does refer to a few previous works, but the obvious emphasis is on genuine field experience of master surgeons, including von Gersdorff himself.

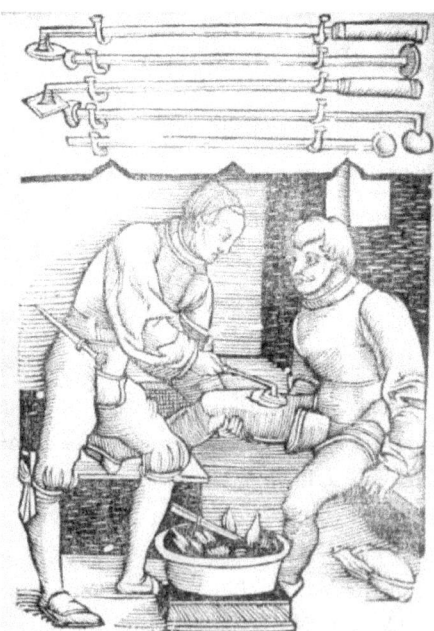

3. Wound widening prior to bullet extraction from a Dutch edition of the von Gersdorff's Feldtbuch, 1651

---

[40] The Woundman image first appeared in the *Fasciculus Medecinae* (Venice, 1492) by Johannes von Ketham and is reproduced with slight variations in the weaponry throughout the century. Guns never appear.
[41] Utterode (1875: 381-84)

von Gersdorff's field manual went through several editions and was translated into Latin and Dutch, but it does not seem to have enjoyed wide acceptance. Its procedure was not the one the young Ambroise Paré observed other surgeons using when he had his first exposure to gunshot wounds during the French king's 1536 campaign to raise the siege of Turin. Paré goes beyond von Gersdorff's accounts of field procedures in describing his reservations about da Vigo's method, which the other surgeons urged upon him.[42]

One time the number of wounded needing oil applications caused the supply of oil to run out, and Paré only dressed the wounds with "egg yolk, rose oil and turpentine." After a troubled night, he returned to find the patients who received only the dressing well rested since their wounds were not inflamed. Those whose wounds were flooded with hot oil were in pain and sleepless. Paré concluded that the hot oil was detrimental to the wound's healing, and that flesh did not inevitably become inflamed due to gunpowder poisoning. He conveyed this discovery to his eminent Paris colleague Sylvius after he returned from the war.

Other surgeons Paré encountered during his wartime assignments agreed that gunshot wounds are best dressed with suppurative medicines that cause the bad matter to pour out and should not be further damaged with hot oil. When the projectile tears through bone and muscle, however, further symptoms difficult to treat arise in the body. If the body is replete with ill humors or in a hot, moist or cold, dry atmosphere the wound will putrify and be difficult to cure. Not the heat of the gunpowder's combustion but "the foulnesse of the patient's body and the unseasonablenesse of the aire" leads to the patient's demise. He gives an example of a Scottish nobleman shot through the thigh with no damage to the bone who was not administered hot oil and fully recovered within 30 days of the wounding. While invoking the ancient humoral account of health and disease in the body, Paré also is using it to adjust treatments according to actual experience.

Paré adopted a rhetoric similar to that of von Gersdorff in relying upon personal experience affirmed by other field surgeons and in rejecting the hot oil treatment of gunshot wounds and the theory of gunpowder poisoning that warranted the practice. He names examples of the success of the less drastic method, and the reason why it may not succeed. Not because of the scalding heat of the gunpowder but because of the humoral condition of the body and the surrounding atmosphere.

To counterbalance devastating intrusions, Paré employed a repertory of extractive instruments specialized for collecting metal and bone fragments, for reducing fractures that have pierced the skin and for removing embedded bullets from bone.[43]

---

[42] Recounted several times over the years in Paré's works. Thomas Johnson's 1634 translation into English, reprinted in Paré (1968: 137-41)
[43] Paré (1575: 371-73)

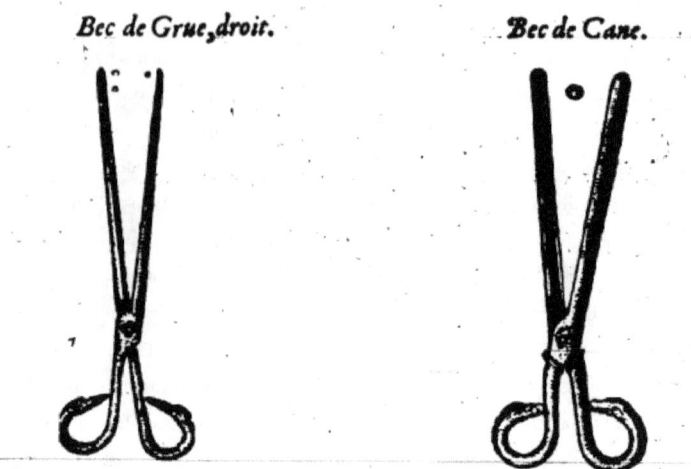

Bec de Grue, droit.   *Bec de Cane.*

4. Scissors (Crow's Beak and Crane's Beak-bird heads modeled in the handles) with toothed blades for picking out fragments of the size shown. Paré (1575: 371)

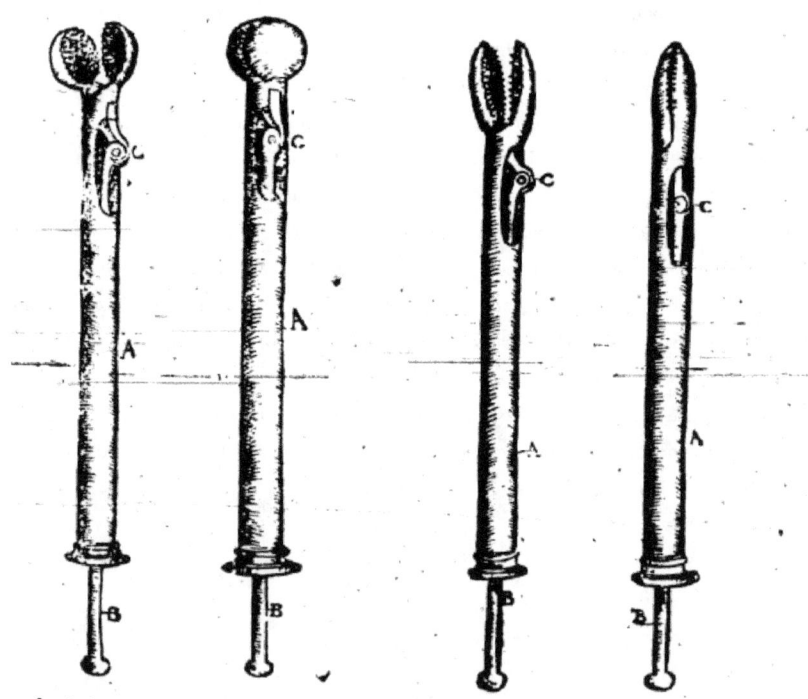

A Monſtre ſa canule. B La verge qui fait ouurir & fermer la charniere.
C La charniere.

5. Bullet extractor called Lizard's Snout for grasping and withdrawing impacted projectiles of different shapes. Paré (1575: 373)

Paré may have administered much less traumatic fluids when first dressing the wound. His armament of tools, and his textual commentary on the occasions of their use, signals a forcible determination to clean foreign matter out of the wound.

At the end of the 16th century battlefield wound surgeons retained the heritage of wound drinks and plasters, mainly to distract the patient. The patient's frame of mind, understood in humoral terms, was an important factor in the success of the surgery and the eventual healing. The serious work of removing the disruptive pieces left by the internal action of the bullet was the main task. Otherwise the body's interior would not return to a balance of humors appropriate to the person.

A bullet that mangled the interior of the body was a grave threat to the order of health. Attention on the part of both soldiers firing the bullets and surgeons pulling them out had shifted, from the surface wound opening upon the poisoned track to the deranged insides. von Gersdorff and Paré published treatment advice discouraging what they considered harmful treatments and announcing the approbation their own procedures received from others working in the field.

Yet even their advances in treatment could not keep up with the mass wounding that resulted from greater distribution of firearms and increases in the force of the bullets. Paré's treatment innovation was born of triage on the battlefield, of not having enough hot oil to pour it into the large number of gunshot wounds. This was during siege warfare, one of the few occasions when the matchlock (mid15th century) and the flintlock (early 17th century) firing systems which required more time between shots than arrows, had any competitive advantage. Archers were a greater force on foot and on horseback, but archers took much longer to train than musketeers, and musket balls pierced armor that resisted arrows.

Puckle's early 18th century patent was in line with the trend of invention. It was a stationary, rapid firing flintlock that would cause a new kind of injury in foreign foes. It was not muzzle loading. Puckle invented a fictive injury as much as a fictive gun. The Turks in the heat of their advance would encounter a barrage of bullets leaving tears and fragments that overheated and disabled the body. Round bullets entered easily and embedded themselves; square bullets bore down with extraordinary force to knock over the furious assailant. The injury was imagined based on existing wounds.

John Ranby, a battlefield surgeon in the train of King George II, set down his "method of treating gunshot wounds" in a brief manual published after his return to England from the Austrian war.[44] Ranby rejected the use of probing instruments to seek out foreign bodies. The wound was very hot and certainly disrupted the body's humoral balance. Ranby bled the victim to release the hot, wet humor, if there already was not considerable blood lost. Aiding the body to recover demanded minimum interference with natural processes.

The mandate to widen a gunshot wound to permit extraction of the projectile was further questioned by John Hunter, whose experience as a surgeon for the military made it clear to him that wound dilation was common practice. "As the dilation of gun-shot wounds is a violence," Hunter wrote[45]

> it will be necessary to consider what relief can be given to the parts or patient
> by such an operation; and whether without it more mischief would ensue; it
> should also be considered what is proper time for dilating.

---

[44] Ranby (1744)
[45] Hunter (1828: 666)

Projectile wounds observed and treated in different parts of the body suggested to Hunter that there was no one principle of treatment to be followed in all cases. The wound was not to be dilated simply because it was a wound, but only if the interior had to be opened for other reasons. Surgeons with knowledge of treatments outside military exigencies would know what else was to be done. There was no universal wound: a gunshot wound to one part of the body differed from a gunshot wound to another part. He continued the prescription of bleeding as the primary constitutional treatment.

The doctrine that surgical probing would exacerbate the wound was beginning to establish itself. The instruments that Paré and his followers used to reshape the wound were going the way of hot oil. At the same time one surgeon or another might revive or modify an older procedure in an attempt to gain an edge on the number and variety of projectiles shot into soldiers in warfare.

Hunter did find that the greater frequency of gunshot wounds posed a challenge to surgeons accustomed to other types of injuries. From the variety of wounds, the projectiles were either lead balls fired by flintlocks or cannonballs and their shrapnel. Hunter also saw the velocity of the projectile as a major factor in the nature of the injuries it caused. Swords and arrows did not enter the body at those speeds. The parameters for the medical study of gunshot wounds were being established, but the sense of what a specific shape and velocity of a bullet could do to a body would long remain disconnected from the pathological anatomy of gunshot wounds. As Hunter himself observed, the experience of surgeons on the battlefield commanded immediate, simplistic solutions without much observation of what bullets actually did in the body.

The percussion cap, which contained a set amount of powder, and then the brass casing, which held both powder and bullet, reduced the amount of time required to load a charge. Engineered rapidity of firing, of loading and reloading, increased the lethality of the bullets packaged in metal cartridges with explosive and primer. The bullet became a component of the cartridge.

The number of treatises on gunshot wounds describing wound variations and treatments increased with the increase in the number of gunshot wounds. Bullets were designed to be made, delivered and fired in quantity. The treatises included the treatment of multiple wounds to the same person, which could be survived because of the nature of the bullets. The increase in velocity, force and numbers was a function of industrial bullet production capacity and fostered a wish for greater efficacy of treatment. Gunshot wounds, like guns themselves, were still mostly a military matter. The London apprentice's playful shot into the street with a militiaman's musket was one marker of a transition to increased civilian use.

The humoral system figured into the treatment of gunshot wounds as it did into the treatment of wounds in general. The wound was an opening into which humors passed and from which humors escaped, causing an imbalance treatment had to repair. A gunshot wound had such an overwhelming effect that the individual's predominant temperament mattered less than immediate issues of repair. The hot oil application that Paré ended, and the cauteries that followed it, were not intended to bring heat to the body; they were intended to constrict blood vessels and stop uncontrolled bleeding.

Surgeons with experience treating gunshot wounds gave pragmatic advice based on their observations of the body's response to the trauma and treatment. Characterizations of temperament figured more strongly in the individual's susceptibility to treatment than in the nature of the wound. Thomas Chevalier, in his 1803 prize treatise on gunshot wounds, recounted the behavior of an impatient elderly man who removed the dressings from his wound to the femoral artery.[46] Chevalier observed the rapid regrowth of vessels visible in the exposed area and restored the dressings. The man's temperament guided his actions and not the response of his body. But his actions threatened the recovery of the body. Gunshot wound victims were growing younger and older and included women.

Surgeons might record that a gunshot patient was of a "bilious" or a "sanguine" temperament as readily as for any other type of patient. G.H.B. Macleod noted of his head wound patients in the Crimean war that an "inflammatory, lymphatic or nervous temperament" made for considerable variation in the ability to recover from a wound.[47]

Thomas Longmore referred to the sanguine temperament of an army lieutenant in his 1862 treatise on gunshot wounds, but in later editions expanded the section on shock to delineate the "variations in the degree of shock" arising in soldiers of different constitutions.[48] He uses language humoral, physiological and psychological to assess the individual body's handling of battlefield energy flux upon wounding.

> A soldier having his thoughts carried away from himself, his whole frame stimulated to the utmost height of excitement by the continued scenes and circumstances of the fight, when he becomes conscious that he is wounded is instantaneously recalled to a sense of personal danger...This alarm, and the depression induced by it, will vary in degree according to individual character and intelligence, state of general health, the structural condition of the heart, whether vigorous or weak, and other personal peculiarities. If the emotion be as intense as it is sudden, it alone will induce a rapidly fatal result in some individuals. In others of a different temperament, the alarm and depression will be controlled, and even extraordinary energy be manifested in its stead for a time.

Longmore goes on to recall the numerous examples of soldiers who have been severely wounded even to the extent of losing an arm, who walk a distance to an infirmary and then collapse. Later when Europeans or Americans saw an Afghan or a Filipino walk a distance after being shot they ascribed it to primitive obduracy that had to be overcome by the bullet itself. Wounded Europeans responded according to a characterology of alarm and depression. The energy that arises in the body of a wounded man causes his death, or propels him to safety and rest depending upon his temperamental and physical characteristics. The humoral has become part of a complex of individual peculiarities deciding the body's final reaction to the injury.

Whether a man who has been shot lives or dies is decided by the dimensions of his personal makeup. Experience had taught surgeons the role of the reaction to shock in the survival and recovery of the gunshot victim. Longmore also examined the effect of individual peculiarities on the severity and duration of shock.

As doctors began to recognize the individuality of soldiers' response to gunshot, they also took account of the variety of bullets shapes entering the soldiers. Longmore expanded a later section of his book to

---

[46] Chevalier (1806: 11)
[47] Macleod (1858: 119)
[48] Longmore (1877: 145-47)

distinguish between three bullet shapes and their separate wounding patterns. Cylindro-ogival and cylindro-conoidal bullets joined the spherical projectiles.

Two soldiers wounded during the same battle in the head of the humerus by differently shaped projectiles gave an opportunity to compare the fractures. A conoidal bullet buried itself in the cancellous bone, caused a web of splits and a fracture extending across the epiphysis laterally down the shaft of the bone.

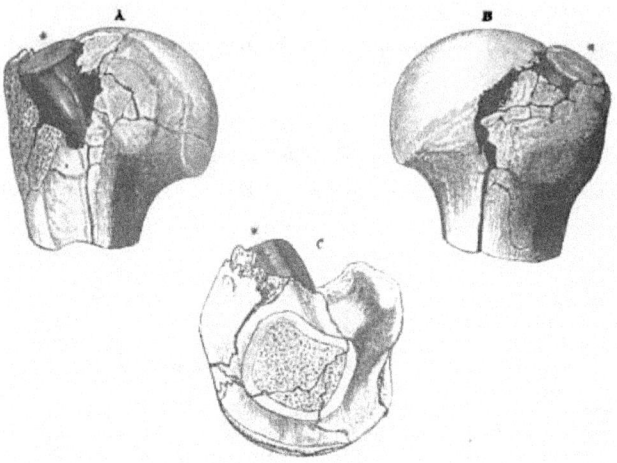

6. A conoidal bullet buried in the head of the humerus. Longmore (1877: 145)

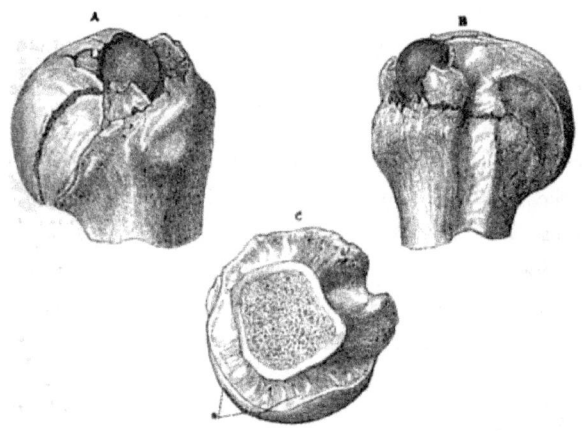

7. A spherical bullet buried in the head of the humerus. Longmore (1877: 149)

The soldiers were around the same age. The victim of the conoidal bullet suffered the graver debility.

Different bullet shapes caused different damage in the same anatomical location. Surgeons began compiling variables of bullet wounding. The treatment of gunshot wounds was becoming layered. The same procedures that applied to all wounds causing bleeding and internal damage still applied to them, but growing collective experience now urged the peculiarity of each wound. There was no common overall wounding pattern and no common treatment for all gunshot wounds.

The variable of temperament, of humoral constitution, was factored in. What was its relationship to the other variables? Did a particular bullet shape affect people of an evident humoral constitution more than others? One aspect of Puckle's gun was an attempt to take advantage of this possibility. There soon was need to explore it further.

4. Big Game Hunting

The European colonial advance into the Americas, Africa and India created opportunities to hunt a greater variety of large animals than at home. The concept of big game hunting emerged as the outsiders displaced their struggle with indigenous people to a struggle with the formidable animals of the region. The buffalo hunt in North America, the elephant, hippo, rhino and lion hunts in central and southern Africa, and the tiger hunt in India had the potential for staging deeds of sportsmanlike conquest. The organized pursuit of animals that individually or in a herd could destroy the hunter but were mastered by his skill and equipment was also the pursuit of administrative control of the territory. The big game hunter could manage the beasts that the human natives merely kept at bay.

As with invasion, conquest and the suppression of revolt, the equipment gave the edge. At first the equipment was guns firing bullets larger and heavier than those needed to kill smaller game and humans. The size of the bullets measured in weight of the shot and then in caliber was considered the decisive element in killing big game. The choice of a larger, heavier bullet traded speed for force imparted by the black powder charge.

A bullet's performance was calculated in a formula that included the mass of the bullet, its effective range, its trajectory and its penetrating power. In close encounters with big game animals, a Romantic ideal for the hunter, a premium was set on the stopping power of the ammunition. The bullet had to prevent the animal from reaching the hunter and from escaping him. Big game that remained mobile after being shot could avoid further shots and die unrecoverable.

The ability of the bullet to inflict a disabling wound was never fully attributed to its greater size. Examination of the remains of elephants or buffalo stopped with one shot gave an anatomical account of stopping power, and confirmed how precisely aimed that shot had to be. Large bullets with short ranges gave little leeway for equipment failure or poor aim however much they permitted a fantasy equation between spear or arrow hunters and gun hunters. The impetus driving invention was the need for bullets with the greatest single-shot stopping power at moderate range. This required an intuitive and then a scientific understanding of the relationship between wound dynamics and gun-bullet design.

The connection between bullet and wound was a connection between shooter and control of the attacking animal.

The effects of bullet structure, propulsion and direction upon the movement of the animal were learned from studying a fatal injury. The physical condition of the bullet remains found in the animal's body bore the traces of the forces that had acted upon it. This was not the same as surgeons gearing treatment of wounded humans to their notions of the nature of gunshots. Game animals were not candidates for healing. The condition of blood vessels, muscles, bone and organs that had been affected by a particular bullet, explained the ease with which the animal was brought down by that particular shot.

I pressed my trigger, and for the first time my carbine missed fire.—Page 211.

8. "The lion killer" confronts a lion (Gérard 1856: 210)

Jules Gérard's assistant during his Algerian desert hunt was one of the eight men who fired when the lion first appeared. Gérard did not have another loaded gun at his disposal after he missed the close shot with his carbine. He had already fired in the direction of the lion's heart as the lion was about to spring, and he had no other bullet to expend as the lion seized one of the men in his claws. Fortunately, the lion fell dead with that one shot. The next day the hunter returned with some men to where the lion lay and carried the lion's carcass to the village in an improvised stretcher. Gérard had the lion's skin taken off, and "observing the effect of my bullets, I abandoned the carcase to the Arabs..."[49]

This encounter with a lion was no accident. The French lieutenant had killed 30 lions between 1844 and 1856. The episode depicted in the Gustave Doré illustration that accompanied the narrative is the template for all the lion encounters, even for all encounters between hunters and big game. The hunter stands forth boldly when the beast is sighted; the natives fire futilely or are absent except as victims. The enraged, still-not-dead beast lunges and is felled by the hunter's force. The hunter examines the remains and reflects on the relative efficacy of his firearms.

Gérard does not name the firearms and bullets he was using, nor does he give details of the bullets in the flesh, only that he examined the skinned body for their effects. The single shot aimed at the heart must be what ultimately finished the lion, though it continued its rampage after receiving the bullet. The Arabs carried the lion remains to their settlement and received them from the hunter minus the trophy skin.

The African lion hunt was aimed at single animals. The bison hunt in North America was directed at animals singled out from a herd by a horseman. The relevant danger was that a wounded bison would turn suddenly and toss both horse and rider. The same principle of not merely angering the animal into a focused

---

[49] Gérard (1856: 210)

attack also applied to this quarry.[50] The decisive shot took on greater urgency when opportunities for multiple shots required balancing and firing the gun with one hand while managing the horse reins with the other.

9. The Bison Hunt. Note skulls on the ground. Davis (1869: 155)

American bison hunters superseded the Plains Indians, whose culture of bison driving became one of bison pursuit after the introduction of the horse. The broad grassy plains at one time supported the greatest number of large land animals in the world, kept numerous by the annual burning of the surface growth to encourage the growth of grasses that sustained the herds. Replacing the Indians' lances, arrows and guns with guns, most Americans did not hunt the bison for sport but as a source of skins, meat and fertilizer for profit. Alarm at the disappearance of the bison multitude was not primarily motivated by concern for species conservation, but by desire to preserve the "masculine frontier culture" represented and described by Buffalo Bill Cody, Theodore Roosevelt and others.[51] Americans seeking actually to hunt big game had to turn elsewhere after the bison herds were decimated.

An English barrister and "forest conservator" in Mysore, C.E. M. Russell, wrote a guide for those undertaking the hunt in India at the end of the 19th century.[52] "Sport," he intoned, "as distinguished from butchery, needs neither apology nor excuse." His book offers advice on sport shooting tigers, elephants,

---

[50] Davis (1869: 155)
[51] Isenberg (2000: 175)
[52] Russell (1900)

rhinoceroses and a variety of smaller game, but it begins with the Indian bison, whose American cousins were indeed being butchered at the time.

Russell considered it an outrage to take aim at female bison. The bull separated from the herd was the only respectable kill. Russell's hunting was on foot, not as close to the animals as a horse and rider could manage. His arms list included the 4 and 8-bore elephant guns used since their introduction in the 1870's and the more recently introduced .577 and .500 caliber express rifles. For bison he recommended a .577 bullet loaded with 6 ½ drams (3/8 ounce) of black powder either "solid" of with "a very small, short hollow filled with a wooden peg." Careful aim and large bullets still might not fell the bull with the first shot, but poor aim and small bullets would leave bullets to accumulate beneath the skin of the unaffected animal. Russell had seen a hide parted from the body of a newly killed animal expose a collection of expended bullets.

During the same period Richard Tjader hunted big game in central Africa with gun and camera. Tjader was an evangelical preacher associated with Dwight Moody who married the daughter of a millionaire and spent several years on the American hunting circuit in Africa.[53] The following sequence of photographs illustrates the discovery of a pair of rhinos first asleep, then alerted to the presence of the humans by their "tick birds," one of them preparing to charge then felled at by a single shot of Tjader's .577 express rifle at 10 yards.

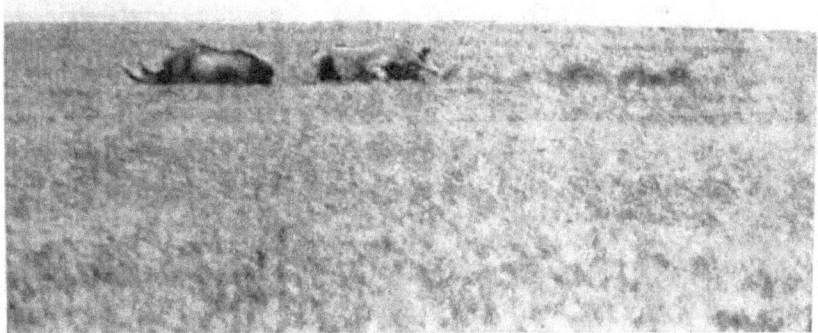

10. Two rhinos asleep on the plains to the northwest of Guaso Narok, distance about forty yards.

---

[53] Big Game Hunter to Preach from Motorcar, *The San Francisco Call* January 25, 1912: 3

11. The same animals. Note the tick birds on the backs of the beasts

12. The same animals. The one facing the camera is about to charge at full speed.

*13. At about ten yards he fell, killed instantly by a single shot from the big .577 express rifle.
Four photograph sequence with captions. Tjader (1910: 132-35)*

The camera was introduced ever more closely into the hunter's confrontation with the dangerous animal, documenting the immediacy of the thrill and danger. Tjader referred at times to the men accompanying him ready to hand him his gun and take the camera when the animals took notice. In one episode not photographed Tjader shot a rhino that suddenly dashed out from the brush and charged to within 6 inches of where he stood before it fell. He recounted the death of an Englishman who had the time to fire twelve bullets from a .500 express rifle before the rhino reached him, killed him and fell dead a few paces further away. Two of the bullets touched the rhino's heart, it was learned from the butchering, and two penetrated its lungs. Tjader declared that rhinos would be one of the last big animals to be completely exterminated.

By the early 20th century the masculine frontier culture of big game hunting, especially in Africa, had reached an edge of reported adventure requiring ever more immediate visualization. The heads of animals were sought for their formidable features and mounted on walls in homes and in museums, such as that of the renowned hunter F.C. Selous, whose collection passed to the Natural History Museum in London.[54] Selous died in January, 1917 when shot in the head by a German sniper while reconnoitering during an African skirmish of the European war.

On the brink of World War I, an experienced hunter named Sinclair was the subject of a story recounted by the Australian adventurer Arnold Wienholt.[55] Having failed in his attempt to be elected prime minister of Australia, Wienholt in 1913 went lion hunting in Northern Rhodesia. He was familiar with the area having fought on the British side during the Boer War, and offered his services to the South African Union to scout for an attack on the contiguous German colony. Sinclair's story was relayed to Wienholt by a Dutch hunter, Rensberg, who with Sinclair conducted a cart carrying the party's belongings as they traveled

---

[54] Millais (1919)
[55] Wienholt (1922: 94-97)

the road. Sinclair and a few Mombakush men came upon a group of lions feasting upon a roan they killed in the road. Sinclair quickly shot one of the lions through the body with his .303 Lee-Metford rifle.

The lions scattered. The lion that was shot moved away more slowly than the others. Sinclair sent a "boy" to retrieve his .470 caliber "elephant gun" that was in the cart a distance behind. Sinclair forgot to give the "boy" the key to the locked cabinet where the gun was kept. He grew impatient waiting. Following the lioness' spoor, he approached her and was again attacked. When Rensberg and the others arrived some time later they found Sinclair recumbent and bleeding with the corpse of the lion beside him. He had killed the animal in a knife fight without discharging his reloaded rifle. His injuries and loss of blood were too grave for survival, and he died soon afterward.

14. The Death of Sinclair, drawing by Walter Seed. Wienholt (1922: 96)

Rensberg joined Wienholt's party and later along the road he asked if they thought a German could be killed with a "hard" bullet: "the point of a .303 bullet, if not a soft nose, wants blunting to kill a buck with certainty." Wienholt assured Rensberg that a German could be killed with an "untampered" bullet. Rensberg then reflected that the German should at least be given a "head shot."

Having witnessed the bloody aftermath of Sinclair's failure to kill the lioness with a "hard" bullet, a .303 that went through her body, Rensberg wondered if such a bullet could kill a beast like a German. A .303 with a blunted point would expand in the body and cause enough internal damage to kill a buck, but that arrangement was not necessary to finish a German. Rensberg then concluded that the German should at least be given the courtesy of a mercy kill.

Armed Europeans in Africa considered it their duty to kill lions when they encountered them. Every day along the road was potentially a lion hunt. Sinclair opened fire with a low caliber service rifle that propelled the bullet through the body of the lion, who survived and moved off. Wanting to be sure of the kill and to take possession of his trophy, Sinclair called for a higher caliber rifle locked away in a wooden cabinet on the supply cart. In the fatal battle with the wounded lion he used only his knife.

Rensberg later equated the lion, which had not died on being shot with a hard .303 bullet, with a German, an enemy in the current war. A .303 bullet filed to expand would have killed the lion but was not needed for a German, who must be dispatched with a hard .303 in a humane manner.

The death of Sinclair sets out the terms of the big game hunt as they had become established by the early 20ᵗʰ century. The small bore service rifles will not kill big game unless the bullets are tampered with to expand in the target's body. Rather than the modified .303, a hunter like Sinclair keeps a large bore rifle on hand to stop and kill enraged animals. He does not keep the rifle close at hand, because the rifle is heavy and Africans who serve as bearers can't be trusted to handle it. Forgetting to give his assistant the key, combined with eagerness for the kill, leads to a fight to the death. Wienholt, who provided himself with a .303 sporting Lee-Metford rifle, a .375 Mannlicher and a .450 express, was well outfitted for animals he might come up against. He does not say that he or any other hunter used expanding bullets, though he clearly recognizes their specific utility. Rensberg's question about the bullet needed to kill an enemy like a wild animal had been asked before.

The high caliber rifle was the basic firearm for anyone expecting to encounter big game in Africa. This type of gun was the center of scenes of firing at a charging animal, like the lion, bison and rhinoceros kills, and its absence led to scenes like the Death of Sinclair. Into this confrontation was introduced the small bore rifle, which was loaded with cartridges containing smokeless powder. The bullets shot from these cartridges maintained a flatter trajectory and thus had greater kinetic energy when they reached the hide of the quarry. This gave them a greater range, and introduced the danger of a perforating wound that passed through the body of the target without hitting a bone or a vital organ. The black powder used in the large bore, high caliber rifles fouled the firing chamber and did not impart the energy to the heavier bullet, which was more likely to cause lethal damage.

Small bore rifles using smokeless powder cartridges were ambivalently adopted for big game hunting as the tradeoff between their advantages and dangers became better known. They were developed for warfare in Europe and served that purpose in other parts of the world, including Africa. What helped make them part of the hunter's kit was knowledge of the effects of bullet modification upon the wound produced by small caliber bullets.

In 1886 the Czech physician Emil Holub, evaluating the best bullets for defense against and hunting of large animals during his travels north of the Zambezi, pronounced long, flattened on one side, partially hollow bullets superior to soft nose bullets. [56] He disdained the soft nose bullets for the wanton damage they caused to the tissues of the animal. Holub assumed that these bullets were for large bore rifles.

A decade later articles in sporting and science journals were advancing the cause of small bore rifles firing expanding bullets with smokeless powder. An 1897 piece in *Scientific American* attributed the greater efficacy of soft point .303 bullets to the force they exerted on "otherwise inert flesh and bone" as they expanded against static resistance. "These bullets, when they strike, spread, cutting and tearing in all directions."[57] The Bengal tiger, the elephant and rhinoceros of Central Africa "are now being successfully

---

[56] Holub (1975: 77)
[57] The Savage Rifle-Smokeless Powder and Expanding Bullets, *Scientific American,* September 18, 1897:181

hunted and slain by sportsmen using the .303 rifle." Casts of wounds caused by expanding bullets propelled by smokeless powder were placed in evidence for readers to judge the extent of damage done by a single bullet. The small caliber bullets if prepared to expand formed wounds as extensive as large caliber bullets without the short range, curving trajectory, corrosive powder and excessive carrying weight of the larger caliber black powder bullets.

15. *Casts of wounds caused by smokeless powder and expanding bullets.* Fn57

Ewart Grogan wrote that his childhood ambitions were "to slay a lion, a rhinoceros, and an elephant, and to see Tanganyika."[58] He accomplished these aims several times over for each animal, and for Tanganyika while making the first south to north traverse of the African continent. Grogan had a decided preference for .303 rifles over large bore rifles. He divided the bullets he used into four classes according to the animals they killed. The solid nickel-covered bullets were best for elephant, rhinoceros and hippopotamus; the lead-nosed stopped small antelope, lions and leopards. The nickel-covered was the least expansive bullet, the least likely to spread upon contact with the animal's body. The lead-nosed were the most expansive. Between the two were two other categories of expanding bullet for buffalo and large antelope.

Grogan analyzes the stopping power and wound pattern of each bullet according to the nature of the animal. For the elephant and other thick skinned animals a fully jacketed bullet that does not expand will enter the hard outer layer and without spreading prematurely on reaching the skin. It has to be aimed precisely at the brain through a vulnerable area at the side of the head. For lions and leopards, likely to attack if wounded by a hard bullet that will just pass through them, an expanding bullet will kill with a first shot. Grogan considers "murdering" lions or leopards from a platform or a shelter "unpardonable, unless inevitable." He has shot a lion in this way, but, he writes, he has also shot a monkey (not the act of a sportsman).

Small antelope also receive the most expansive bullets to halt them before they bound away. The larger antelope and buffalo are more susceptible to carefully aimed bullets with less expansive properties.

---

[58] Grogan and Sharp (1902: xv)

Grogan does not dismiss the large bore firearms. He has worked out a schedule of animal targets and bullets that gives him the sport he seeks with small bore rifles. The small bore outfit is the sporting standard in the present day with expanding bullets and solids used according to the animal hunted. [59] The confrontation that a large bore gun permits a hunter expertly aiming a high caliber black powder bullet at an advancing lion remains the backdrop fantasy of big game hunting. The parameters of the hunt are large and small bore rifles, solid and expanding bullets, the anatomy of the animal, especially the thickness of skin and the wound expected.

During the early 20th century, as Europeans in Africa at war with each other hurled newspaper accusations of expanding bullet use contrary to the Hague Protocols, hunters were testing out the comparative worth of expanding bullets for the quick kill. The Berkeley tennis champion Drummond MacGavin, accompanying a mining expedition in South Africa, took part in what he called "the slaughter of the hippopotami," 14 in all, in the Kapue River.[60] MacGavin explains that the thick layers surrounding the animal's body makes the best shot one aimed directly into the head. The theory that only solid bullets can penetrate the hide and skull of the hippo was disproven. The expedition leader killed a hippo with one soft nose bullet that, it was later discovered, passed through skull and brain without leaving a trace of lead and only stopped on the other side of the skull. Butchering the hippo immediately revealed the success of the single shot soft nose fired into a hippo, though others required several bullets before they went still.

Theodore Roosevelt, on safari at Lake Naivasha in what was then the British colony of East Africa, sat in the bow of a boat (alongside the animal) and fired a single soft nose bullet into the back of a hippopotamus. The naturalist Warrington Dawson, who accompanied the expedition, wrote[61]

The monster quivered, rippling the water, and then, with a mighty groan, turned over and floated, as is shown in the picture. The photograph was made only seconds after the fatal shot was fired.

16. Theodore Roosevelt's hippo kill. Fn 61

The still mass of the hippo framed by some papyrus plants is testimony to the finality of the expanding bullet and the dominance of the ex-President over formidable African wildlife.

---

[59] E.g. Brown (2012: 63-67)
[60] MacGavin (1906)
[61] Here Are Hippos That Teddy Shot, *The Seattle Star* September 4, 1909: 10

Coincidentally printed beside the article on Roosevelt's hippo hunt was a brief news notice on a Chinese cook at a California ranch shot dead by a deputy sheriff who had come to arrest him for putting poison in the food of farm hands. There is no indication what type of bullet the deputy fired.

The single shot soft nose felled large beasts wherever they were hunted: a rhinoceros in India[62] or a bear in Alaska.[63] The designation "soft nose" encompassed a growing variety of designs. Bullet makers offered changes in structure that they claimed would make the shot expand uniformly. Leslie Taylor of the Westley Richards company in England fashioned a "capped" soft nose that guaranteed uniform expansion. Early in the century, however, hunters did not note in their dispatches and travelogues what type of soft nose they used. Safaris were outfitted with a supply of both soft nose and solid bullets. A hunter's gun with a multi-shot magazine might be loaded with several different types in alternation.

For all the reports of dangerous animals subdued with a well-placed shot at the last second, there also were reports of a soft nose bullet dropping one animal immediately while an identical bullet, causing an anatomically identical wound in another animal in the same herd, did not prevent it from fleeing.[64] The performance of the bullets, apparently including the improved designs, was irregular and unpredictable.

Ernest Hemingway, on safari during the 1930's, loaded solids and not soft nose bullets for large animals.[65] Bullets that disintegrated before penetrating would infuriate the animal and endanger the hunter. Toward the end of the century Robin Hurt maintained the same caution when hunting buffalo in Africa, until he was introduced to a new formulation of the soft nose that did successfully bring down the massive animals.[66] An experienced hunter has learned discretion, and an innovation has to be convincing in practice not just in theory and test. The bullet must change shape in the right way.

The hunter's courting of risk to gain an individual triumph calls for a refined knowledge of gun, bullet and animal parameters. That knowledge contributes to bullet design or at least is addressed when promoting a new bullet configuration.

Having migrated to Africa while still a teenager, probably to escape the dangers of being a Protestant minister's son in Ireland in the 1920's, John "Pondoro" Taylor acquired enough experience tracking and shooting big game animals to compose a lively encyclopedia of the rifles and cartridges best for that purpose. Taylor wrote in exile from South Africa, expelled because of his anti-apartheid views and his homosexuality. He formulated an index of the Knock-out Value (KOV), the ability to decisively halt an animal, ascribed to a range of bullets and the guns that fired them, from .240 caliber (low end of the scale) to .600. Taylor did not consider the small bores to be "safe weapons to take against dangerous game at close quarters in thick cover, altho a few men used them in the past with considerable success."[67] The weight of the bullet counted more than muzzle pressure or muzzle velocity, and was the one factor that increased in direct proportion to KOV.

---

[62] Hunting the Indian Rhinoceros A Gigantic Beast, *The Queenslander* (Brisbane, Australia)January 12, 1924: 2
[63] Mallinckrodt (1922: 57)
[64] Antelope hunting in New Mexico. Cottar (1915)
[65] Calabi (2011)
[66] Hurt (2006)
[67] Taylor (1948: 14)

Bullets fired at high velocity did have the ability to "pole-axe" a large animal, but only if the bullet was massive enough not to fall apart. Taylor speculated that the fluid shock caused by the bullet entering the hide jarred the nerves of the animal. The Ross .280 rifle introduced to Africa around 1910 brought that "peculiar killing property" to notice. The little, light 140-grain expanding bullet the rifle fired disintegrated if it happened to hit a bone and dissipated its killing power before hitting a vital point. The rifle was rejected "after one or two men were killed by lions which their .280s failed to stop." Taylor favored expanding bullets, soft noses and hollow noses, only in calibers large enough for a high inherent KOV. He related hunting anecdotes to support his figures.

Taylor's views and equipment have of course been subject to commentary and revision. He did state the conditions of the personal hunt as one man against nature. Many of the animals he so skillfully overcame were won without licenses. His autobiography proclaims him an ivory hunter but he was in fact an ivory poacher.

Present-day contexts of animal procurement that are not governed by the big game hunting ethos still reflect its gun-bullet-wound ecology but are not limited by personal confrontation strategies. The Congolese soldiers and their agents who supply the elephant meat and ivory trade use a low caliber machine gun, the AK-47, which can kill an adult elephant with a spray of 60+ rounds without anatomically sensitive aim.[68] An alternative is a single shot to the brain or a vital organ with a 12-gauge shotgun firing a bullet formed of many pieces of shot welded together and fitted into the same casing. The accuracy of the sportsmanlike shot is not needed when the meat and finally just the ivory of the elephant are the only object.

The present mass slaughter of elephants in Africa aims only at procuring the ivory for the Chinese market in turn to pay for weaponry for more wars.[69] The skin and meat and even the ivory of young elephants is left behind by the militarized shooters. The Romantic big game hunter used superior weaponry to defend himself, the colonial enterprise and the "natives" from marauding beasts. Despite the weaponry he might be reduced to a raw knife-against-claw fight. Now the weaponry is discharged, not by those who consider themselves hunters, but by merchants firing into the animals from a safe distance, even from helicopters, to gain foreign exchange. The hunt then is for the human enemy in the context of the arms trade.

During the 1890's, as smokeless powder-propelled low caliber expanding bullets were coming into greater use among hunters of big game, the cultural frame of this technology expanded. The rampaging animals defended against began to include people who needed to be stopped as they attacked the defenders of civilization.

---

[68] Stiles (2011: 31)
[69] Gettleman (2012)

5. Flat Head Bullet for a Ghazi

During the Second Afghan War (1879-80) invading British India forces were beset by ghazis, Muslim warriors who entered camps singly and did as much damage as possible before being subdued. Their seeming indifference to the weapons of the soldiers became a matter of record.[70]

> In 1879 Major Edin Baker reported:-"I saw Captain H., of the Bengal Cavalry,
> empty five shots from his revolver into the back of a Ghazi, who was running amuck
> through camp, at less than five yards' range, without stopping him. I examined the
> man myself afterwards, and found the marks of all six bullets in his body. I consider
> the service revolver should throw a heavy ball of .5 inch to .55 inch diameter, and I am
> half inclined to believe a flat head to the bullet would be an advantage."
> The English army revolver at the time was .455 caliber, shooting a cartridge
> containing eighteen grains of powder and a 250 grain bullet.

The heavy, large caliber bullets perforated the man without causing him to fall, much as they at times did not stop an attacking animal. They were fired from composite cartridges with primer and a strong powder load. The bullets fired at the ghazi had rounded heads. Major Baker is given to consider that a flat headed bullet would effective in stopping the man as it was in stopping an animal.

The revolver attained the rapid firing sequence that Puckle projected for his Defence. In calling for flat head bullets Major Baker was not calling for square bullets. A range of bullet nose shapes had been in use for a while.

The American army officer Alfred Mordecai prepared a report on European military technology based on his visits to European military installations from 1855 to 1856. The report, published at the brink of the American Civil War, included descriptions and diagrams of firearms and ammunition. The array of bullets in use among the nations, in this time before the wide adoption of composite cartridges, ranged from flat to rounded to ogival to pointed heads and internal cavity configurations for the powder.

---

[70] Gould (1894: 202)

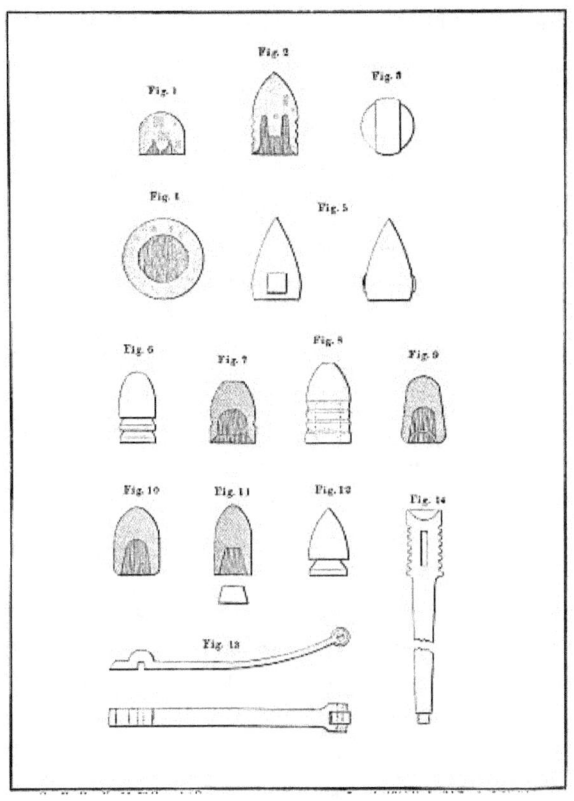

17. Profiles of bullets in use in Europe 1855-56. Mordecai (1861: Plate 21)
Figures 13 and 14 show rammers.

The composite cartridges that had come into use for handguns mounted these bullets in a casing loaded with charge and primer. The rounded or pointed noses afforded less wind resistance and could be aimed more accurately at a greater distance than the flat noses which were more likely to yaw off course. On the other hand the sharper the bullet nose the faster it moved through intervening media, including flesh. Flat nose bullets in the era of black powder firearms were known to have a curved rather than a straight trajectory. They might not have enough energy on reaching the target to penetrate unless fired at close range. This was known from their use in hunting.

The introduction of smaller calibers and smokeless powder that gave the bullet greater velocity out of the muzzle placed the sharp nose and flat nose extremes in an uncomfortable alternation. The lever action rifle introduced in the 1890's required flat nose bullets to be loaded in its magazine because a pointed head might set off the primer of the bullet in front of it.

During the propaganda wars of the early 20th century one enemy accused the other of sharpening the acceptable flat nose (Germans against the English[71]) or of flattening the acceptable sharp nose (Russians against the Poles[72]) exclaiming that each was barbaric and in violation of international agreements.

Hybridization between the two extremes might lead to bullets adapted to different conditions. Baker's flat nose recommendation reached into the history of bullets to find one peculiarly adapted to the circumstances at hand. Who was the ghazi to need a flat head bullet to stop him?

The word "ghazi" is derived from an Arabic root meaning "to strive, to attain" which later became specialized for holy warriors and political leaders. The mutual cattle raiding among pastoralists, between and within social groups, was led by individuals with the daring to enter the camp and lead off the cattle without surrendering to the raided group. Stealth was counterbalanced by bravado. The ghazi's individual initiative was striving toward a success that benefited his kin and himself as part of his social group.

This aspiration was translated from material gain to victory over non-Muslims. Carrying off possessions and animals, or destroying them when pillage was not possible, was a strategy in a worldview that divided humanity into Muslims, those who had surrendered to Allah, and infidels, who were attacked, robbed and harassed until they sought the security of being within the sphere of Islam.

"Ghazi" was a title adopted by, and given to rulers of Muslim states. The title identified them as advancing the boundaries of Islam by the same kind of conquest Mohammed had encouraged when he directed Bedouin tribes people to stop raiding each other and together raid the infidels. The word and the character of the ghazi spread to all peoples converted to Islam. The founding Turkic ruler of the Ghaznavid dynasty of Persia and Central Asia, Mahmud of Ghazna, established a ghazi persona.[73] The 15th-16th century first emperor of the Mughal dynasty in India, Babur, also of Turkic descent, rejected his non-Muslim ancestor Tamerlane and remade himself into a holy warrior.

Osman I, the 13th century founder of the Ottoman Empire, was known as "Gazi" (Romanized Turkish spelling) for his leading warrior refugees from the Mongol conquests in attacks upon the provinces of the declining Byzantine Empire.[74] A Turkish epic poem listed the attributes of the ideal gazi, the instrument of the true religion, and attached the title to Osman and his predecessor for their deeds.[75] The history of the Ottoman Empire was the spread of Islam through military action sanctioned as a drive for conversion that encompassed the lands, properties and resources of the conquered peoples. These were the Turks whom European Christians armed themselves against.

As the Ottoman Empire also declined the agency of the ghazi surged into the realm of struggle wherever infidels approached. The name of the founder of the Turkish republic, Mustafa Kemal, was prefixed by the general honorific "Gazi" and followed by the special honorific "Ataturk." The submarine previously commissioned as the USS Diablo was renamed the PNS Ghazi upon being transferred to the Pakistan Navy in 1961. Mahmood, the young Afghan soldier who shot and killed two American trainers at a shared base in

---

[71] England Manufactures Sharp-Pointed Bullets, *Los Angeles Times* August 4, 1907: 6
[72] Czar Continues to Win Battles, *Ogden Standard* August 31, 1914: 10
[73] Anooshahr (2009)
[74] Cleveland and Bunton (2009: 38)
[75] Lowry (2003: 18-19)

Afghanistan, escaped to join the Taliban and was given the title "the Ghazi of Ghaziabad"[76] in circulated videos.

The proliferation of the ghazi title over the centuries as a marker of royal self-image as conquerors of infidels and defenders of Islam preserved and extended the making of individual ghazis. Troops from British India invading Afghanistan in the First and Second Anglo-Afghan wars encountered fierce Muslim fighters who were called ghazis. Denying that genuine ghazis ever were numerous, the journalist Howard Hensman wrote an account of how ghazis were made. They were peasant men fired by the advance of an infidel army to take up arms against the invaders.[77]

> The *moolah* is to these ignorant peasants the link between this world and the next; in him they place all trust, and as they listen to his fierce harangues, they are ready to do all that he requires of them. He is vested with mysterious attributes; and with quiet assurances promises that "if they attack the infidels in the proper spirit and in full faith, "bullets shall turn harmlessly aside, bayonets shall not pierce them, and their poshteens [cloaks] thrown over the cannon's mouth shall check shot and shell.

Hensman relates these promises to the rousing words of all false prophets. For the Muslim peasants the ghazis who survived several perforating rounds to enter the lines of the better armed defenders confirmed the power of faith over guns even if the bullets did not bounce off. For soldiers like Major Baker the toughness of the attackers was the result of bullet mechanics and not faith. While not noticing how this persistence after serial wounding reinforced faith for the defenders, Major Baker did introduce another strain of intercultural experience into the ghazi's attack when he wrote that the man "ran amuck."

The word "amuck" or "amok" had entered English from the Malay language through travelers' accounts of men suddenly and without immediate provocation murderously assaulting others. In colloquial English it refers to an individual who goes on a rampage, but the men named "amouco" in the Portuguese chronicle of Gaspar Correia (written between 1512 and 1561) were loyal subjects of the king of Calicut (southeast India) who sought revenge against an enemy king for the death of their princes in battle.[78] The amouco, "according to their customs," shaved off all of their body hair, parted from their family and consigned themselves to death charging enemy soldiers and wreaking havoc among the general population until killed.

Other instances of this collective "martial amok" were set down by 16th and 17th century Portuguese and Dutch writers on the Malabar coast of India, Malaya, and Indonesia. The early 16th century also saw the first use of the word for solitary men who went amok, and either were killed in self-defense by their intended victims or captured and judicially executed.

Solitary amok runners continued to appear in these areas after the martial variety no longer was observed. The individual engaged in wild acts of indiscriminate slaughter became the image attached to the word "amok" by the later English colonialists in the region. Alfred Russell Wallace's account of amok as a tactical strategy in *The Malay Archipelago* (1869) was possibly the last witnessed instance of martial amok in any

---

[76] Rosenberg (2013)
[77] Hensman (1882: 335-36)
[78] Spores (1988: 12)

language. By that time the British treatment of solitary amok as a form of insanity to be handled by the courts, and more likely to lead to institutionalization rather than to execution, coincided with the decrease in reports of that behavior.

Major Baker saw the ghazi acting as a solitary amok. He applied the behavioral generality the word was acquiring to the specifics of warfare during the Afghan war. The ghazi was a person and amuck described his action. Physical indifference to wounds by weapons was seldom part of amok advances in the histories, and when it was it took the form of "stopping neither at fire or swords" and running deliberately onto the points of swords.[79] Amok runners do not continue their course after gunshot wounds, but neither did ghazis until they were one of the elements in the encounter between colonial troops and fighters against foreign occupation during the late 19th-early 20th century. Swords, knives and daggers were the equipment of the amok. With guns they were something else.

Amok has been grouped with ghazi and other "war neuroses of primitive societies" such as the berserking of ancient Germans and Scandinavians and the juramentado of the Moro in the Philippines.[80]

Berserk, like amok, has been drawn into general usage in English to denote a behavior, not a type of person. The berserker in the original sense of the word never faced guns. The juramentado did.

18. Attack of the juramentados. Montano (1886: 145)

During his 1879-80 voyage in the Philippines the French physician and ethnologist Joseph Montano had encounters with juramentados in the Islamized southern islands. The closest came on the island of Sulu, when he found himself standing in a suddenly deserted marketplace. Eleven juramentados had entered the village carrying sacks of animal feed and upon a prearranged signal dropped their burdens and attacked anyone at hand with their krisses (curved swords).[81]

> At this moment a woman dashes up all disheveled, followed by a filthy Suluan,
> pale as a newborn. He holds in his hand a kriss dripping with blood. The woman

[79] John Nieuhoff found this in his travels among the Nairs of Malabar. Quoted by Spores (1988: 15-16)
[80] Devereux (2000: 268)
[81] Montano (1886: 144)

cries to me: los juramentados! And like a cannon shot topples me over as she
runs past. Two gunshots go over my head. I get up and see the juramentado struck
in the chest, but he rises too and throws himself, kriss raised, at the soldiers.
Pierced by a bayonet he remains standing, trying to reach the soldier who keeps
him at the end of his rifle. The other soldier reloads his gun and finishes this fanatic
(enragé) once and for all.

Montano then reviews the damage: 11 juramentados dead, but 15 soldiers also, and many others horribly wounded. The juramentados were armed only with krisses. An "amouco," driven by an insult, still straining forward with a spear run through his body was described by Diego de Couto in the seventeenth century.[82]

The lowlands of Sulu were dominated by a sultanate challenged by the Spanish imperialists who had colonized the northern islands. The Muslims and then the Christians attempted to convert and dominate the peoples living in the hilly central region of the island. The guns in the hands of the soldiers and the word "juramentado," meaning someone who has taken an oath to kill the enemies of Islam, were Spanish introductions. Juramentado behavior preceded the arrival of the Europeans, who in this case applied their own term rather than assimilating an indigenous word, which probably was "manuju."

The Moros and other Muslims distinguished between amok, a sudden outburst of deadly violence, and juramentado, which had a religious inspiration and direction. The organized violence of the juramentados, moving in a group, gaining entrance through a ruse and charging as individuals, resembles military amok.

Like the ghazis contemporaneously attacking British Indian troops in the Afghan provinces, the juramentados brandished blade weapons against gun-bearing soldiers. A native weaponry confronted imported firearms: Montano wrote that the soldiers fired "remingtons," which only can have been the Remington Rolling Block .43 caliber rifle standard in the Spanish army 1870-93. These were breech-loading single shot rifles. The high casualty rate among the soldiers was due to the limited stopping power both of bullets and bayonets when meeting the onrush of a sword-swinging enemy. The soldier who held the juramentado at muzzle distance in Montano's account survived only because his companion could reload and fatally shoot the transfixed man. The other juramentados were stopped only by a squad of soldiers firing at them simultaneously as they approached. Volley fire against the hand weapons of the natives once again.

The difference between blade and bullet, between indigenous and colonial warfare, was bridged by the rage of the warrior. The ghazis and juramentados, who were crafty in gaining the element of surprise, also seem to have known that the bullets would run right through them, and would not stop them. Nor would the slender bayonets, before they could cause severe injuries with rapidly swung broad blades. The ghazis and juramentados did not have firearms at first. The attackers did not disarm the soldiers they killed and use the impressive-looking rifles because the attackers were much more deadly with their blades. The soldiers fell because they relied on their static weapons.

Montano adds two pieces of information to the juramentado dossier when he glimpses how unclean the attacker was and how pale he was. "Sale comme un peigne," he wrote, idiomatically comparing the man's body state to a comb run through wool and picking up quantities of stems, leaves, soil and feces from the sheep's exposed coat. This implies that the man had been outside for a long time communing with the earth

---

[82] Yule and Burnell (1968: 21)

and forest without attention to his appearance, unlike others in this society. "Tellement pâle qu'il paraît verte," "so pale he appears green," in the sense that a newborn is "green." Montano's photographs of Sulu people, men and women, laborers and royalty, show all of them to be dark-skinned.[83] The juramentado's paleness as of a newborn whose skin has yet to darken must have been striking.

The juramentado is feral. He is careless about his contact with the earth and surrounding nature, like an animal. He is bloodlessly pale. His cold skin does not respond to painful piercing and stabbing. He is not boiling with hot blood; he is in a cool rage.

Religious fanaticism may characterize the juramentado's final drive, but it does not explain why he has consigned himself to the frenzy leading to certain death. Some sources attribute the forming of a juramentado to the draconian traditional law of the island of Sulu, which makes a debtor and his family the slaves of the creditor. Swearing the oath is the only way to avoid this existence.[84] Juramentados then separate themselves from the rest of society (shaving off hair, saying fairwells to kin) and become members of a suicidal attack force at the service of the sultan. The colonialists' histories of encounters with ghazis and juramentados emphasize their religion-driven fierceness, their ability to approach armed soldiers, which justifies aiming concentrated fire and uniquely powerful projectiles against sword-bearing adversaries.

America won hegemony over the Philippines in the 1898 Spanish-American war, and with that susceptibility of its troops to attacks by juramentados. The Americans took up the existing Spanish attempt to subdue the Muslims and "traditionalists" of the southern islands. They also extended the account of the Moros and the juramentados. Tales of their wild persistence in the attack began to appear in the reports of American field officers. General Samuel Sumner, commandant of the Mindanao district, included a translation of a Spanish-language article on the fighters in his report to Congress for 1903.[85] The chief candidates for juramentado oathing were "fame-thirsting youths" who sought a noble death, and relief from creditors, and whose inflamed passions made them invulnerable to the bullets of the soldiers.[86]

Included in the 1903 military report was a history of juramentado attacks during the period of Spanish suzerainty from 1878 onwards, and notices of a number of incursions by single and multiple juramentados against American troops and their Filipino allies. Several of these accounts were by Captain John J. Pershing, who over the following years applied his earlier experience fighting Indians in America to campaigns against the Moros.

Pershing reconnoitered his adversaries culturally to detect points of vulnerability. While it is not clear whether he or another officer came up with the deterrent of burying killed juramentados with dead pigs to render them too unclean for entry into a blissful afterlife, he came to recognize that religious passion was not the sole key to the recruitment of candidates for the pledge to die killing Christians. A decade later the then Brigadier General Pershing explained in his annual report as military governor of the Moro province that

---

[83] Not appearing in his published work but archived by the Bibliotheque Nationale.

[84] Varigny (18??: 189-90). A 15th century source, without naming the frenzy, makes the person who kills a raging debtor responsible for his arrears. Yule and Burnett (1902: 20)

[85] U.S. Congress. House of Representatives (1903: 303-4)

[86] A "juramentada," a female juramentado, is never heard of. Retana (1921: 110)

young men of modest means were unable to compete with village chiefs (datus) for young brides and consequently threw themselves into the martyrdom to be gained in war.[87]

Pershing observed one incident that dictated a change in the arms carried by U.S. soldiers. An officer discharged all the chambers of his .38 Colt revolver into a juramentado and was still sliced to death by the man's barung (a razor sharp short sword) before another soldier nearby killed the attacker with a blast from his .45 caliber firearm. This led Pershing to call to replace the weak .38's with the older Army weapon of the Indian wars, the Colt Model 1873 Army revolver, also known as "the Peacemaker" from the quiet that usually followed its loud discharge. Other officers citing similar incidents with juramentados and insurrectionaries in the Philippines made the same demand.[88]

In 1892 the Army had begun to replace the single-action .45 Colt revolver in service since 1872 with a double action .38, but that was before troops encountered the close and sudden fighting conditions with blade swinging opponents that the Philippine war introduced. While the old .45's were brought back into service (and some apparently never left), gunmakers were invited to submit proposals for a new design that would combine multiple shots with a heavy caliber. John M. Browning's Colt model 1906 semi-automatic with a detachable 7-round magazine holding 255-grain .45 caliber bullets was adopted as the service pistol in 1911 and remained the chief Army hand arm until 1985.[89]

Nasser Marohomalic, a Maranao (Moro) member of the United Nations Commission on Human Rights, credited "the martial character and fighting prowess" of the Moros with bringing about the change in U.S. Army arms and regulations.[90] The negative publicity in the American press generated by 1906 massacre of Moro men, women and children seeking refuge in a volcanic crater (Bud Dajo) by U.S. troops under the command of General Leonard Wood also was a factor in bringing about the change. This battle had begun with an amok by a Moro named Pala who resisted capture by General Wood and spurred local resistance to the American mandate. General John J. Pershing, who took a more circumspect approach to the holdouts on Bud Dajo, later led an assault on a Moro position (Bud Bagsak, 1913) using the new Colt .45's rather than heavy ordnance and obtained a victory less distressing to the American public.[91]

The numbers of dead juramentados produced by the American efforts yielded opportunities to advance the description that had begun with Montano. Arthur Henry Savage Landor, painter-explorer-ethnologist on an American-sponsored visit to Sulu, also witnessed an attack on a marketplace by juramentados who had concealed weapons. He did not get a glimpse of the men as they drew their burungs as Montano did. He was allowed to make an anthropometric examination of the corpses of three of them halted by American bullets. He cast their physiognomy in terms of "criminal types."[92]

> These men had square faces, very flattened skulls, and low foreheads, cheek bones low down in the face, and so prominent that when in profile they nearly hid the excessively flat noses; weak and small receding chins, and the square-fingered, stumpy , repulsive-looking hands typical of criminals-as

---

[87] Pershing (1913: 71)

[88] Vidmer (1905: 187-88) places this call in the context of Army service pistol history

[89] Stewart (2005: 367)

[90] Marohomsalic (2001: 21)

[91] Fulton (2011)

[92] Savage Landor (1904: 1,238)

cruel hands and heads as I have ever examined, the animal qualities being extraordinarily developed. Their repulsive appearance was also somewhat enhanced by the hair of the head being shaved clean, the mustache and eyelashes removed so as to leave a mere horizontal strip of black hair. The teeth had freshly filed and stained black; the hair of the arm-pits pulled out, and the nails of the fingers and toes trimmed very short.

Savage Landor tabulates measurements of arm, leg and head that he believes will allow readers interested in criminology to compare these men to others with similar measurements and inclinations. For criminal anthropologists body appearance, measurements and shape of skull belied mentality and innate tendencies of a brutish human type resembling animals. Criminal metrics were a cross-race constant. The practice of depilating the head, familiar from other observations of juramentados and also known as a ritualized warrior initiation in other parts of the world, is joined by teeth-filing and staining, which was a cosmetic practice of both men and women among the Moro and was not related to status as a juramentado. Elsewhere in his book Savage Landor refers to the teeth-filing typical of another group, the Subanaos.

As was often true in the researches of those who try to affirm absolute behavioral categories associated with specific body measures, Savage Landor did not compare the juramentados he examined to others who were not juramentados, nor to others who were. He confirmed an assumed correspondence of appearance with mentality based on a very small sample.

In his 1924 travel memoir *Everywhere* Savage Landor recounted bringing up Pershing's name in a conversation with President Theodore Roosevelt, who at first did not know how to spell it. Savage Landor had accompanied Pershing on an expedition, and was acquainted with his approach to indigenous peoples. His recommendation helped Pershing gain the favor of a president who shared the view of savages behind the army officer's tactics.

There was evidence that the resistance to bullets had more to do with the nature of the projectile than with the temper of the target. On October 26, 1905 Antonio Caspi attempted to escape imprisonment on the Philippine Island of Samar.[93] An officer shot him 4 times with a .38 Colt revolver but Caspi continued his flight until knocked out with a blow to the forehead from the butt of a Springfield rifle. Evidently the soldier handling the rifle quickly concluded that this was the end of the gun to subdue Caspi. One bullet passed through the man's torso, another through his arm, and two lodged in his back. None of them did internal damage so critical that Caspi lost the impetus to escape.

---

[93] Le Wald (1907: 192)

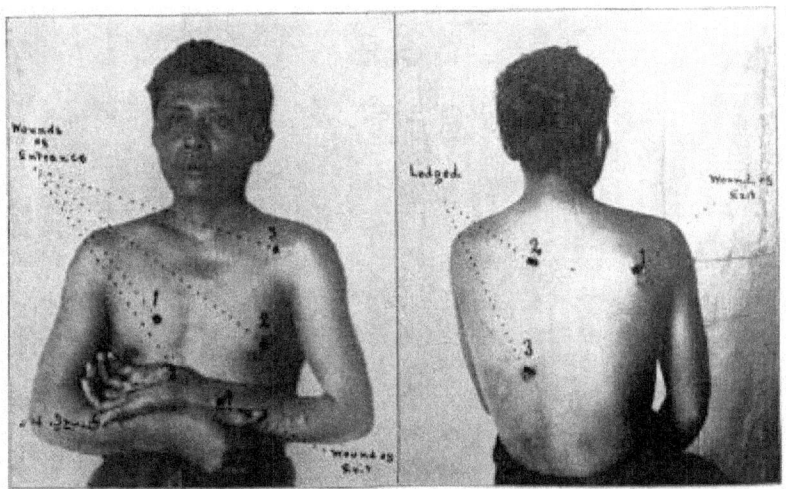

19. Antonio Caspi: location of .38 caliber gunshot wounds. Le Wald (1907: 193)

Caspi was not a juramentado. He was fleeing prison when shot, not attacking American troops. He was a captain in the pulajanes, a mixed group of brigands and followers of a Christian prophet on the island of Samar, who raided representatives of any central authority attempting to assert itself on the island.[94] The pulajanes attacked with a short, broad-bladed knife called a "bolo."

Caspi's survival was not used as an instance of pulejane invincibility, but as a guide to the use of firearms to disable opponents. It was possible to fire a .38 bullet multiple times into a man and not cause him to lose the power of independent movement. Four random shots into Caspi at close range didn't accomplish this. Examples like that of Caspi, and experiments with cadavers informed physicians that it was necessary to use a bullet of sufficient cross-section and to aim the gun to strike a vital organ or break a bone of locomotion.[95] Juramentado or any other fighter, frenzied or calm and calculating, had the same body and could be stopped in the same way. Louis La Garde, the best-informed medical observer, noted that the American military in the Philippines had chosen a .45 caliber Colt revolver firing steel jacketed bullets for both practical and diplomatic reasons.

Ranged against the forces of foreign socioeconomic domination, in Africa, South Asia and the Philippines were individuals and small groups who could resist the small caliber bullets being adopted by the industrially supplied armies of the Western nations. This threatened to demonstrate that the weapons of the invading powers were not superior to inspired bodily force. For the invaders, however, it was a matter of technology that would allow them to maintain a defensive stance. Clearly a change in ammunition was required to halt the attackers cold. The flat head bullet, or a reversion to the higher caliber formerly used against the American Indians, might prove sufficiently destructive. The solution reached at an outpost of the British

---

[94] U.S. War Department (1905: 89-92)
[95] La Garde (1916: 69-71)

Empire in India, an adaptation of a big game hunter's technology to defend against other humans, would provide a template or at least a name for the decisive projectile.

6. The Dum-Dum Complex

A French physician who attended to German wounded during World War I afterwards recounted a visit from a German colleague.[96]

> A German doctor who visited the ward of the German wounded, on seeing one
> of the terrible wounds produced by regular bullets but bullets which, under
> certain conditions, produce wounds as of an explosive effect, said to me
> [in German], "That is a Dum-Dum bullet." I answered him [in German], "No,
> that is the explosive action of a regular bullet." Well might I insist on the German
> word "regular"; he had his opinion and I had mine.

The physician, writing in French, recalls the conversation in German so there is no mistake about the exact word (Dum-Dum) the German physician used. He cannot convince the man that the wounds are not the result of the Allied forces using dumdum bullets. Ordinary bullets, that is .38 caliber fully jacketed bullets, can tear apart flesh and bone like the forbidden bullets. The book in which this conversation is printed gives evidence that the Austrians, Axis combatants, were the ones using dumdum bullets. No conciliation in the telling of the disagreement, only firm accusation in one direction.

Dutertre-Dulévíéleuse, the French physician in the exchange above, had his own preoccupation, with explosive bullets. He recognized the explosive effect of regular bullets caused by a discharge of the arriving bullet's kinetic energy in tissue offering relatively greater uniform resistance. This effect, and that of dumdum bullets, both were distinct from the effect of bullets containing a charge designed to explode in the body. Dutertre presented evidence that the Austrians also were producing bullets that did explode in the body of the victim, a violation of an international agreement that preceded the one outlawing expansive bullets. From the Austro-Hungarian invasion of Serbia that precipitated the war, the Allies and Axis powers were united in discovering exploding and dumdum bullet atrocities.[97]

These accounts enfolded the previous century's development of ever deadlier weapons and the attempts to formulate legal restrictions limiting them. In atrocity accounts the exploding bullets came to the fore, but they were giving ground to the dumdums, which had the advantage of being far less determinate in their effects than the genuine exploding bullets. The European powers, through the dumdums, imported the boundaries between civilized and savage of their colonial wars, and colonial atrocities, into Europe.

Dumdum bullet wounds preceded the dumdum bullets made to cause them. Dumdums are truly bullets take shape across cultural and political boundaries, and causing the flesh they reach to take shape as well. The American authorities explaining their revised armament in the Philippines were careful to exclude their bullets from the dumdum conformation. Of course they also were testing dumdum bullets there, and faced their participation in the dumdum complex in their own border war with Mexico. Dumdums began with the need to stop the colonized from overwhelming the colonizers.

---

[96] Dutertre-Dulévíéleuse (1916: 6)
[97] Reiss (1916) documents with numbered circumstantial accounts, diagrams of the bullets and photographs of victims the numerous atrocities of the Austrians in Serbia alone.

The phrase "stopping power of bullets" codified the conditions that presented themselves in the colonial enterprise: troops brought from a distant homeland and together with recruited locals facing fighters attempting to recover their own homeland. It was best to equip these defenders of the Empire with the means of securely stopping "the natives." Whether equipped or not with firearms of their own, the natives could not only threaten the lives of the troops by resisting bullets fired into them. They could then continue to advance upon troops unprepared for hand to hand combat.

The anonymous author of a note in a military surgeons' journal put it succinctly:[98]

> But war has demonstrated that the stopping power of bullets depends less upon the
> anatomical lesion than upon the morale of the mass pushing onward with the determination
> of winning the battle.

Which suggests that the morale of the defenders also was at issue. Steps were taken to make the stopping power of bullets the decisive factor, because that at least could be mechanically ensured.

This need for stopping power converged on places like the Royal Artillery armory at Dum Dum, a small town in the vicinity of Kolkata (Calcutta). The name "Dum Dum", not an uncommon one, means either "a mound" or "a Gypsy encampment." A transportation center on the periphery of a vast city (now an airport, railroad station, and last stop on the Kolkata metro line), in the mid-19th century Dum Dum was a zone of transition between the maritime city and the plains to the north, a place from which the material assertion of force over old imperial domains could be launched.

The armory had been the site of a contention over bullets in 1857, the year of the "Sepoy Mutiny." One element in the uprising of soldiers recruited from Indian socioreligious groups was the rumor that the paper cartridge containing bullet and powder for their muzzle-loading Enfield rifles had been greased with pork or beef tallow. It was expedient to tear open the paper with the teeth, which would put Muslim and Hindu troops in contact with animal matter forbidden to them by religious law. This polluting act would cause the Hindu soldiers to lose caste. Soldiers were court martialed for their refusal to use the bullets.

The trend in small arms development in the years after the Mutiny was toward smaller caliber bullets capable of greater velocities, which created the risk of the lead body of the bullet fouling the rifle bore. Repeat action rifles, the Lee-Metford and its successors in the British domains, were engineered for smaller, lighter bullets numbers of which could be carried in the rifle's magazine. A succession of shots was the proposed discouragement of wildly advancing hordes since Puckle's time. The old large caliber balls and lead bullets, slow to reload through the muzzle as they were, had the advantage of flattening upon impact with flesh, though they didn't have the kinetic energy to penetrate far enough to do damage.

The Lee-Metford rifle firing a .303 cartridge was the standard issue for British troops by the mid-1890's. The .303 cartridge, a lead core graded cylindrical shell fully jacketed in a cupro-nickel casing, at first was a black powder bullet. The metal jacket kept the lead bullet out of contact with the rifle bore and maintained velocity that would have been hampered by the easily melted lead. The pressures for more rapidly firing and more reliable firearms induced by conflicts worldwide led to the introduction of the Lee-Enfield rifle and

---

[98] Gunshot Wounds in Manchuria, *Journal of the Association of Military Surgeons of the United States* 17(1905): 217-18.

smokeless powder cartridges. The new rifles could be loaded more quickly and the smokeless powder did not corrode the gun barrel as black powder did.

The soldiers stationed at Dum Dum carried Lee-Metford rifles armed with .303 cartridges when in 1895 they answered orders to intervene in a succession dispute between heirs to the rulership of a small state, Chitral (present-day north Pakistan).The British forces successfully installed the friendly candidate on the throne and in the process extended the writ of the Raj. Encounters at close quarters with fighters for the opposing party encouraged doubts about the effectiveness of the arms deployed against this latest group of charging fanatics.

20. The Chitral Expedition. Gardyne lithograph from a sketch. Illustrated London News,
June 6, 1895

In this illustration the sword and rifle carrying attackers take rifle and pistol fire from advancing British soldiers. A fortified hut appears amid the trees.

One of the "Chitralees" who walked miles to a British field hospital for treatment of his injuries was found to have been shot in five different places with .303 cartridges without the desired effect of stopping him completely.[99] The .577 Snider bullets and the .577/450 Martini-Henry bullets seemed to stop the foe more effectively than the .303's. Worse still, the old .45 caliber rifles carried by a few of adversaries killed British soldiers. The expedition may have achieved its political aims, yet it left the soldiers and their officers with a feeling they shared with their counterparts elsewhere, that the conditions of combat with the weapons at their disposal exposed them to the possibility of being slashed to death no matter how many shots were put into the attackers.

In the aftermath of the Chitral Expedition, the Adjutant General of India Gerald de Courcy Morton, demanded an improvement in field performance. Someone, possibly Captain Neville Bertie-Clay, the supervisor of the Dum Dum arsenal, devised a counterpoise to the threat to defensive colonial soldiering posed by the raging warrior. The severe Metford rifling of the small bore gun barrel and its Lee bolt action were sending the bullet powered by smokeless powder at too high a velocity to stop the enemy.[100]

Bertie-Williams' plan was to file the copper -nickel jacketing away from the nose of the Mark II .303 cartridge to expose the lead bullet. Prior experience with exposed lead core projectiles for animal hunting assured the users of the improvised bullets that the head of the bullet would "set up" or expand upon making contact with the flesh of the target, cut a wide channel through the flesh and shatter any bone it reached. The prepared .303 "dumdum" bullets combined low caliber speed and lightness with high caliber damage to flesh due to projectile flattening. The low caliber bullet became higher caliber entering the body with the added kinetic energy of the lighter bullet propelled by the same charge.

The Dum Dum arsenal was an ammunition factory, and making dumdums from .303 cartridges was a technique of bullet remanufacture. The dumdum innovation was to turn mass-produced low caliber fully-jacketed bullets into an equally large number of dumdums. The manual technique was easily taught.

The next British imperial campaign against indigenous people, the Tirah Expedition in 1897-98, became known for the explicit use of dumdum bullets against warriors stirred by a religious leader. The Afridi people who had been subsidized by the British to guard approaches to the Khyber Pass suddenly rose up and commandeered the forts. In the public narrative the new bullets gave the British and Indian soldiers defending civilized order the battlefield edge against charging raiders.[101]

This narrative was perpetuated by a small number of official sources reporting the distant events. Several campaign histories by military officers and journalists complicated the triumphalist rhetoric. The British expeditionary army was not the only force to fire dumdum bullets at the enemy: there were numerous

---

[99] Hamilton (1898: 12)
[100] Spiers (1973: 3-4)
[101] Appleton (1899: 334), for example

instances of British troops wounded by the same bullets.[102] Either a trade in dumdum bullets had spread the ammunition or the Afridi and their allies had learned how to dumdum bullets.

Familiarity with those wounded by dumdum bullets raised doubts about the bullets' vaunted stopping power. One commentator related the story of an Indian soldier who was hit five times and continued fighting, but then affirmed that at close ranges the dumdum bullets have considerable force.[103] Field surgeons who had examined the wounded during the Tirah campaign questioned even this advantage.[104]

> These cases and others which came to my notice during the campaign lead me to believe that the dum-dum bullet at close range certainly does not set up, if only soft tissues are traversed; and that even if a massive bone is struck the resulting injury, though very serious, will compare not unfavorably as the regards the possibility of recovery with similar wounds caused by Martini-Henry or Snider rifles.

These named rifles were the old standards firing .458 and .577 caliber bullets respectively. By emphasizing the equal chance of a victim's recovery the writer, a surgical officer named C.M.T. Thompson, both questioned the relative worth of the dumdum bullet and as a surgeon dispelled suspicions that it was inhumanely injurious. That word "humane" was entering into the public discourse on bullet capabilities with the deliberations of The Hague Conference in 1899.

The official disclosure that British army troops were armed with dumdum bullets during the Tirah campaign was intended to boost national pride. The arms trade in the bullets and rifles and the informed doubts about the stopping power and range of the bullets were not part of this promotion. The government also would have preferred not to have it known that army officers and surgeons had conducted executions of captive mullahs, Muslim religious leaders, using .303 and .457 caliber bullets in order to compare the effects of the different calibers on their bodies. [105]

The first major campaign in which improvised dumdum bullets were used was the punitive expedition that set forth in 1896 to avenge the 1885 death of General Charles George Gordon, and to win the Sudan from the forces of the successor to the Mahdi Muhammad Ahmad, who had defeated and beheaded Gordon. Gordon had been sent to evacuate British civilians from the Sudan but, as much a religion-inspired leader as his opponent, contrary to orders precipitated a siege of the garrison at Khartoum. The British government vacillated in making plans for the punitive expedition. The forces under General Sir Herbert Kitchener attacked the Mahdists whose most fervent fighters were the Ansar, called "dervishes" by their opponents. The term "ghazi" was not used except by way of comparison in the literature on the campaign. Hindu troops from British India and Muslim troops from Egypt and the Sudan were in Kitchener's force.

Kitchener, or at least those in his army, defeated a much larger army than his own forces in a battle near Omdurman, the Mahdist capital close to Khartoum, because of superior firearms and the ammunition used. Ever more rapid communications meant that newspapers and journals were able to report events in the

---

[102] Lionel James (1898: 173, 198, 241, 245, 255)
[103] Carless (1898: 88)
[104] Thompson (1898: 14)
[105] Lawrence James (1998: 410)

Sudan soon after they occurred. In its April 23, 1898 issue the London illustrated newspaper *The Graphic* printed this lithograph by a staff artist from a sketch made by an officer on site.

21. Making Dum-Dum Bullets at Damarli, near Berber. The Graphic, April 23, 1898

Entitled "Making Dum-Dum Bullets at Damarli, near Berber," the print shows the men of the Royal Warwickshire regiment systematically removing service bullets (.303 Mark II) from boxes on the right and passing them to the men seated legs extended beneath a course of planks in the center. The men are holding steady a file with the right hand as they shave the metal jacket off the head of the bullets held in the left. The dumdum bullets are boxed in the same containers by the row on the left. Before they are reboxed an officer gauges each bullet to make sure they have not lost too much diameter in the shaving, making them loose in the gunbarrel. The operation shown yielded 8 boxes of dumdum bullets every two and a half hours, or 60 cartridges a minute. The illustration kept the public abreast of the weapons being made to punish Gordon's murderers.

Damarli was the summer camp for British forces about 30 miles down the Nile River from Omdurman, where the decisive, but not the final battle would take place on September 2, 1898. Over 9 thousand Mahdists killed against fewer than 50 British troops lost argued persuasively for the weapons and method of projectile preparation. Soon factory-like improvisation passed into production as the Mark III bullet and its successors were manufactured with a fully jacketed body and a soft point or a hollow point lead nose that would collapse and mushroom upon impact.

Omdurman also generated a windfall in fine "athletic" skeletons of dervishes that were marketed by London dealers in human bones. [106] The fallen warriors supplied the demand for remains superior to the stunted bones of paupers and executed criminals more readily available.

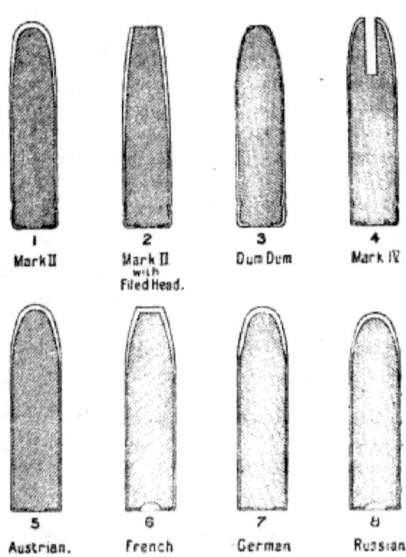

22. Bullet profiles in 1899, by nation. The top row is the British contribution. The filed Mark II and Dum Dum are not jacketed at point while the French flat nose is. The Quarterly Review 190 (1899): 161

A note in a journal for ordnance specialists indignantly denies that the dumdum bullet ever was the service bullet for British forces.[107] Col. Bainbridge, the head of the Royal Laboratory at Woolwich Arsenal near London devised the Mark IV bullet, .303 caliber round manufactured at the arsenal with a slot across its jacketed nose. The open space would cause the head of the bullet to mushroom on impact but it was not correct to group it with the shaved head jacket dumdum bullets. The dumdum category had come to include any bullet visibly engineered to expand upon contact with flesh, whether jacket shaved at the apex, made with lead nose exposed, with a slotted or a hollow point. Field tests comparing the soft nose dumdum with the Mark IV gave more favorable ratings to the Mark IV in set up and range.[108]

An 1897 article in *Scientific American* was careful to use the phrase "expanding bullet" to name the ammunition combined with a new rifle design and smokeless powder to kill the most formidable big game in

---

[106] London's Skeleton Market, *The Times* (Washington, DC) December 11, 1898: 1. An article printed with variations in American and British newspapers for the next two years.

[107] Service Bullets, *Arms and Explosives* 7(1899): 6

[108] Review article, *The Quarterly Review* 190 (1899): 159-60

Africa and India (elephants, Bengal tigers).[109] Casts of the bullet's wound tracks photographed for the article (Illustration 15) were labeled "wounds caused by expanding bullets and smokeless powder" without reference to the targets they came from. "These bullets, when they strike, spread, cutting and tearing in all directions," the article concluded. The expectations surrounding the use of the bullet were apparent; they were the same expectations guiding the manufacture of dumdum bullets in the war against the Mahdists. Their flesh would be torn apart.

Those examining the course of the bullets in flesh took notice when the new low caliber bullets and firearms were introduced in the 1880's. Studies of the relationship between the morphology of the gunshot wound and the force and composition of the projectile had grown in sophistication and become an experimental as well as an observational science, eventually under the heading of wound ballistics. The field executions of the mullahs at Omdurman were one of these experiments.

A U.S. Navy Surgeon, Henry Beyer, gave a lecture at the Naval Academy in 1894 reviewing research on the wounds produced by the low caliber bullets. Most of the studies he considered, in several languages, were experimental in nature: the results of firing the bullets into inert materials such wood and clay, into vessels filled with liquid and into living and dead animals and human cadavers. He recognized that the damage to bone, the most readily photographed, ranged from penetration to cataclysmic destruction. An English army captain had, for instance, test-fired Lee-Metford .303 rounds into horses at stated ranges and making note of the appearance of the exposed bone.

23. Low caliber gunshot wound of horse bone shaft. Beyer (1894: Plate II, Photo A)

---

[109] The Savage Rifle-Smokeless Powder and Expanding Bullets, *Scientific American,* September 18, 1897: 181

This demobilizing, possibly fatal destruction of the long bone shaft was the same type of injury that Dutertre-Dulévièleuse twenty years later tried to convince his German colleague was caused by conventional bullets. Under other circumstances the picture would be far less daunting. Beyer attributed the variability in the results of these bullet experiments to the different projectiles used by the experimenters and the different conceptions of what constituted an explosive effect.[110] Often the projectiles were assessed comparatively by nation.

24. Dumdum bullets: left unfired, middle fired into distant target, right fired into near target.
Nimier and Laval (1899: 61)

A French study of the wounding action of projectiles employed by European militaries at the end of the 19th century lists and illustrates in profile drawings the cartridges mounted with bullets of 17 nations. Modeling of the bullet nose and cartridge surface was nationalized, a matter of style as much as function though practical reasons were given for most features. According to the authors the British were not "satisfied" with the performance of dumdums at Tirah, and loaded manufactured hollow points for the expedition against the Mahdi in the Sudan.[111] Drawings of dumdums expended at long and close range were given in support of the performance record, followed by drawings of expended bullets from test firings corresponding to the national types.

This lack of a scientific standard of bullet performance further blurred the focus when the already vague category of dumdum bullets was refracted through national interests. Paul von Bruns, a Tübingen physician who earlier had published several experimental studies of low caliber bullets listed by Beyer, in the late 1890's experimentally reproduced the wounds caused by bullets fired in the overseas wars. The title of one paper placed "Dum-Dum Geschosse," "Dum-Dum bullets," in parentheses after "Bleispitzengeschosse," "Lead-pointed Bullets." Another study published the same year (1898) simply referred to the "Inhumane Kriegsgeschosse," "Inhuman Ordnance." These papers were in one way the continuation of the existing wound ballistics studies. A number of other papers in German, English, French, Italian and Russian reported

---

[110] Beyer (1894: 159)
[111] Nimier and Laval (1899: 61)

similar investigations of the yet newer firearms and bullets.[112] Several of these papers also used the label "dumdum" to apply to a variety of preparations.

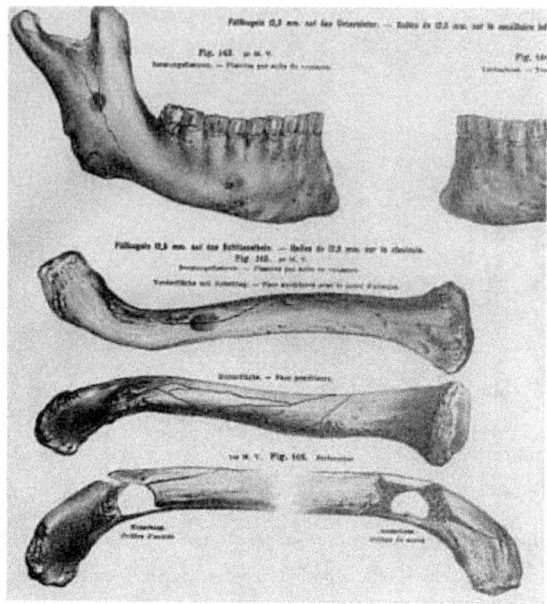

25. Dumdum damage to mandible and clavicles. Von Bruns (1899)

In the plain universe of medical studies von Bruns' use of the accusatory word "inhuman" to refer to the Mark IV bullet manufactured by the British touched a raw nerve. von Bruns' findings led to the conclusion that bullets like those being made and used by the British caused undue pain and likelihood of death in warfare. It was as if the decades of death and destruction caused by explosive shells and incendiary bombs, fired by cannons and before long to be deployed from the air, in both internecine and overseas conflicts could be placed outside the microscopic frame of the dumdum bullet expanding in viscera.

An English surgeon who had written on the surgical impact of the dumdum bullet, Alex Ogston, observed that von Bruns' experiments on dumdum bullets employed Mauser big game bullets which had a much greater portion of the lead core exposed.[113] Providing translated portions of von Bruns' paper target Mark IV experiments as well, Ogston appealed for further experiments before the dumdum bullets were banned. He brought Beyer's trials of a new U.S. Navy rifle into evidence as well, in support of the difficulty distinguishing wounds caused by conventional low caliber bullets impelled by smokeless powder and similarly impelled dumdum bullets.

---

[112] U.S. War Department. Office of the Surgeon-General. (1916: v. 21, 296-97) indexes many studies
[113] Ogston (1899a)

This continuing lack of a scientific relationship between bullet design and the consequent wound ballistics despite many efforts to gain that knowledge, and the lack of an agreement about which bullets were dumdum bullets were offset by a need for something that could do what a dumdum bullet was supposed to do to the body of a determined, probably indigenous, opponent. Dumdum bullets had stopped those opponents in a few encounters. These lacks and needs formed a complex labeled "dumdum bullets." The bullets' continuously taking shape was a permanent and necessary feature of their longed-for definition.

The boundaries the bullet was aimed across were national and racial and themselves shifting with the shape of the bullet. British involvement in the Western response to the "Boxer rebellion" in China inspired a statement from an apologist for the dumdum typology.[114]

> Bullets of an expanding character, although prohibited by the laws governing war among civilized nations, are freely employed in cases where the foe is of a dusky hue, and while the English have refrained from using the dumdum bullet in South Africa, owing to the fact that the enemy by whom they were confronted was white, like themselves, there is no doubt whatsoever that they will use them in China, just as they did throughout the Tirah campaign and throughout all the Indian frontier troubles. Indeed, the Indian troops that have gone to China are equipped with no other ammunition than these dumdum bullets. Then, too, the circumstance that England should dispatch Indian instead of white regiments to China indicated that the latter is regarded as being "beyond the pale," so far as obligations of the rules of war are concerned.

In the absence of civilized warfare those of a "dusky hue" can also be marshaled by the civilized to let fly dumdum bullets at others not within the pale of civilized warfare. They had already done so on the Indian frontier and in the Sudan, and they did again in China.

After these foreign troops raised the siege of European legations by the "Boxers" and the Chinese Imperial Army, together with foreign civilians they looted Chinese palaces and residences.

Dumdum bullets were indeed fired in that campaign. Frank Richards, an English private who served in the British army in India for seven years with a firm belief in the mandate of empire happened to be present when British troops entered the home of an elderly Chinese man in search of treasure rumored to be hidden there.[115] One of the white British soldiers, Richards recounted in a memoir published decades later, wanted to bayonet the man when he kept denying there was hidden treasure in his house, but his companion was curious about the effects of dumdum bullets on the human body. Experimentally blowing off the back of the man's head did not bring the hidden treasure any closer.

Though this dumdum murder was committed by a white soldier aiming across cultural lines, the sepoys or Indian troops who formed the bulk of the British army in and from India, also took part in the looting. A letter from the Australian adventurer George Ernest Morrison, to Ignatius Valentine Chirol, the director of the foreign department of *The Times of London* for which Morrison was Peking correspondent, traces another line of sight.[116] Despite the efforts of the dishonest commander of the British Expeditionary Force to

---

[114] Cleveland (1900: 589-90)
[115] Richards (1936: 142)
[116] Morrison (1986: 1, 194-95)

suppress the details, Morrison learned that a sepoy had "run amok" in the legation, shot one German and wounded two others, and that the dead sepoy's bullet case was full of dumdums. Another sepoy found with dumdums was brought in and charged with possession of unlawful ammunition in order to placate enraged German officials and avoid further incidents.

Whether disingenuousness or genuine ignorance was behind treating sepoy possession of dumdums as a misdemeanor, the dumdum shooting of Germans by a British soldier of Indian origin and the move to manipulate the consequent report was a line across boundaries and into the future.

7. Bullets That Expand or Flatten Easily

The 1899 Hague Peace Conference opened on May 19, the birthday of Tsar Nicholas II of Russia. The Conference had been proposed by the Tsar and his foreign minister Count Muravyov the previous year. The agreements reached were signed by the attending delegates on July 29 of that year, its four sections and three additional declarations to go into force on September 2, 1900. The sections established a Permanent Court of Arbitration for the peaceful settlement of international disputes, and set a number of rules for the conduct of war on land and at sea. The declarations placed limits on the launching of missiles and projectiles containing asphyxiating gases. Several of the agreements reaffirmed and extended earlier conventions.

Declaration III, "On the Use of Bullets that Expand or Flatten Easily in the Human Body" makes reference to a convention initiated by a previous tsar. It was declared to be "inspired by the sentiments which found expression in the Declaration of St. Petersburg of the 29th November (11th December), 1868." The alternate dates are due to the misalignment of the Gregorian calendar in most of Europe with the Julian calendar in the Russian Empire.

"The sentiments which found expression in the Declaration of St. Petersburg" were the conviction that projectiles that could explode and flame within the body were contrary to the laws of humanity. The Russians had been in the forefront of developing such projectiles, which their troops had fired against European adversaries during the 1853-56 Crimean War.[117] A desire not to receive such missiles in return, in addition to their inhumanity, motivated the call for a negotiated ban, which the Declaration phrased as a renunciation.

The Contracting Parties engage mutually to renounce, in case of war among themselves, the employment by their military or naval troops of any projectile of a weight below 400 grammes, which is either explosive or charged with fulminating or inflammable substances.

This agreement did not cover explosive shells or missiles launched against troop concentrations, ships or settlements. The Russian diplomats explained that smaller explosive projectiles were meant to destroy munitions stores and not people. The size limitation forestalled the firing of explosive bullets at individuals.

The Declarations of the 1899 Convention were intended to limit the application of some developing technologies to warfare. Dropping bombs or projectiles from balloons was prohibited, as were projectiles that spread asphyxiating or poisonous gases. The third 1899 declaration was an extension both of the sentiment and the technical parameters of the 1868 Declaration.

The Contracting Parties agree to abstain from the use of bullets which expand or flatten easily in the human body, such as bullets with a hard envelope which does not entirely cover the core, or is pierced with incisions.

---

[117] Packard (1882: 445-46)

The word "dumdum" did not appear in the language of the Declaration. Initial shape and final shape were stipulated. Soft nose bullets and bullets with incised noses were banned but not bullets with a hollow in the nose. Exploding bullets made by putting contact explosives in the hollowed out interior of a conventional bullet already were prohibited under the 1868 Declaration.

Incised bullets specifically were prohibited because they could be made spontaneously from any of the bullet types in use. Cutting lines into the head of a bullet was a practice of hunters and soldiers difficult to monitor. The delegates anticipated the carryover of this practice into warfare.[118]

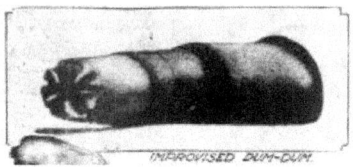

26. Improvised dumdum bullet.

A story that made the rounds of American newspapers in 1896[119] illuminates common alterations of bullets to be expected of soldiers in the field, to which the protocols were addressed. A New York sport hunter shot a small deer that ran away and lost itself in the forest. Four years later the hunter was talking to a woodsman who hunted in the same locale. The woodsman told him that he had found a flat nose bullet with the marks of a knife blade in its nose lying on a rock shelf not far from where the deer had been shot. The city man recognized the bullet as one he had prepared for the hunt. The deer must have died in the forest and been dismembered by animals. One of them, probably a fox, carried a share of the remains to the table-like rock. The incised flat nose was only recognizable because it hadn't expanded. Hunters, and soldiers, cut their bullets as a matter of course.

Soft nose bullets whether flat or curved could be incised. Most of the delegates considered hollow nose bullets to be exploding bullets, one stage in the evolution of dumdum bullets, even when they were labeled "non-explosive."[120] The 1899 Declaration attempted to cover all forms of bullets that changed shape in the body, any of which might explode. They would expand violently and release particles whether they contained explosives or not. Expanding bullets, exploding bullets, and dumdum bullets all merged in perception.

Declaration III, like its St. Petersburg predecessor, was an agreement among nations potentially at war with each other to forego deadly anti-personnel weapons. An opt-out clause was attached to the Declaration. Any party to the agreement could state one year in advance that they were no longer abiding by it. There was no prohibition against stockpiling expanding bullets in anticipation of the other party using them first.

The British delegates, joined by the Americans, cast the only votes against the Declaration. The Portuguese delegate abstained. The British charged that the ban on expanding bullets was engineered by the

[118] Dum Dum Bullets as Seen by the Scientist, *The Washington Herald* December 6, 1914: 6
[119] Story of a Bullet, *The Princeton Union* (Princeton, Minn.) September 17, 1896: 7
[120] Hamilton (1898)

Russians to interfere with British military action in Africa. Ogston's paper published in the *British Medical Journal* soon after the Conference concluded was dedicated to lightening the onus of the negative British vote, to make British bullets appear not so uniquely barbarous. It included comparisons of bullet profiles, stories of damage and photographs comparing dumdum wounds with wounds by ordinary low caliber bullets.

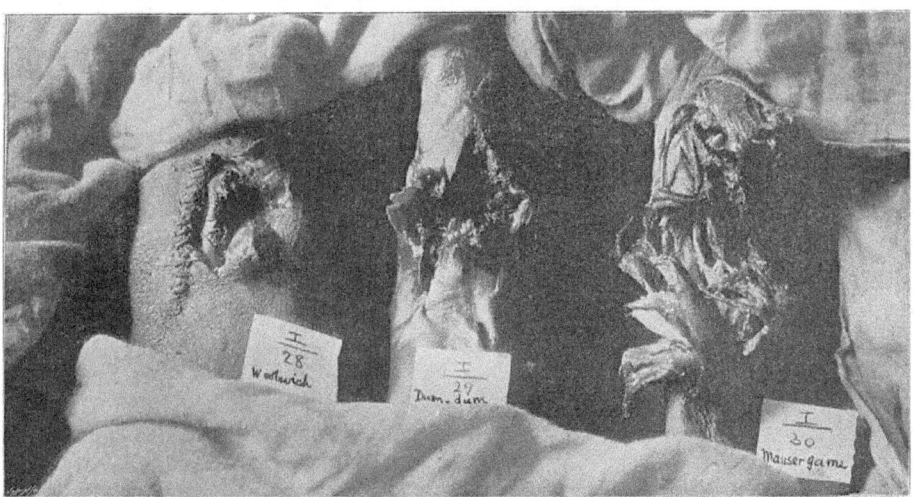

27. Woolwich, Dumdum and Mauser game bullet wounds to arm bones compared. Ogston (1899b: Fig. 4)

Other photos compare the bone destruction done by the same type of bullet arriving at different angles. And the degrees of destruction caused by different bullets. The Declaration's expanding and explosive bullet category could apply to almost any bullet, and might not apply to dumdum bullets.

The Americans had already turned to the destructiveness of bullets of a higher caliber than service bullets, though they preserved the option of exploding rounds. The British, soon followed by the Russians, were engaged in military action that formed the post-Declaration world of bullet exchanges and accusations.

British imperial control of South Africa was at first assisted then greatly hampered by the presence of Dutch, German and Huguenot settlers known collectively as Boers ("farmers," in Dutch or Afrikaans, the local language based on Dutch). An 1880-81 war had ended in a truce between the British and the two Boer polities, the Transvaal and the Orange Free State. The discovery of large, easily mined gold deposits made another play for control economically feasible for the empire and precipitated another war beginning in 1899.

In the course of combat British doctors discovered "hideous lacerated wounds" in imperial soldiers: evidence that the Boers were in possession of dumdum bullets. Ernest Blake Knox, a physician serving with the Natal Field Force under British command, wrote in his account of the 1900 campaign that the Boers captured thousands of rounds of dumdum bullets when they took the town of Dundee.[121]

Knox excused the British from "uncivilized" intent in possessing the bullets, saying they were brought by Indian troops but then abandoned because the commander prohibited their use and ordered that no more dumdum bullets be landed. Knox's section on the bullets' presence was in part in response to a writer supporting the Boers who accused the British of having them in the first place. These bullets did not fit the Mauser rifles of the Boers, who could make their own expanding bullets cause the same damage. Knox had found in Boer camps soft nose, slot nose and dent nose bullets, mistakenly identified as "explosive" in the press.

The equivocation surrounding the British supply of expanding bullets was brought forward in a "semi-official" War Office report summarized in the newspapers.[122] Running short of cartridges for their Lee-Metford rifles, commanders were allowed to access the ten million Mark IV bullets in storage, most of them already in South Africa. Troops in England about to deploy to South Africa were each issued 50 Mark IV bullets which they were to use exclusively for practice. The flagrant violation of international law, especially by commanders who had complained about Boer use of expanding bullets, did not escape public notice.

On the other hand a military magazine published in the Netherlands, where there was strong support for the Boers, published a photograph of cartridge wrappers marked "Dum-Dum" found on a battlefield, which led to a discussion of whether these were the allowed Mark II bullets or the forbidden Mark IV bullets.[123] Dutch medical personnel returning from South Africa recounted maiming wounds that could only be due to British dumdums. A pro-Boer diplomat quizzed returning Red Cross doctors to extract eyewitness testimony of unmistakable expanding bullet injuries without much success.

All the elements of the post-Declaration dumdum complex are present here: the exchange of accusations and denial of their use, the horror of the wounds, the multiplicity of types, easy self-manufacture and the vexed belief that they are explosive. Added to these is the refinement that the bullets the British had but didn't use were brought to Africa by Indian army troops, possibly even from Dum Dum itself. Furthermore, Dum Dum was a manufactory and not all bullet packages labeled "Dum Dum" contained dumdum bullets.[124]

The backdrop to all further interchanges of accusations on the subject of dumdum bullets is that they have been declared "uncivilized", thus not for use in "civilized" warfare which qualified them for use against "uncivilized" people who had their own horrendous means of dispatching enemies. The Boers, Europe-derived, were not in the category of the uncivilized. They had their own guns, could make their own bullets and had their own X-ray equipment to visualize the internal damage. The damage the Boers wreaked upon British forces was in part due to the accuracy of the Boers' marksmanship delivering the bullets in a way that maximized their potential. The dumdum accusation was a way of sidestepping the marksmanship issue as the long war continued.

---

[121] Knox (1902: 227-28)
[122] British Troops to Use Expanding Bullets, *The San Francisco Call,* January 12, 1900:1
[123] Kuitenbrouwer (2012: 207)
[124] Use of Expanding Bullets, *The New York Times,* March 10, 1900

Questions asked during a session of the House of Commons Committee of Public Accounts on May 15, 1901 traced the ins and outs of dumdum bullets.[125] The Committee was considering compensation to contractors whose bullets supplied to the army turned out to be defective not from faults in the manufacture but because the specifications given them were wrong. The bullets delivered had to be withdrawn "not because of wounding but because they fouled rifles."

The bullet separated from the casing prematurely and left a lead deposit in the barrel of the gun, ruining the gun for further firing. The Committee had to be implicitly reassured that a mechanical problem and not the Hague Declaration was the reason the bullets were removed from supply and either destroyed or used for practice. "Not because they wound too much but because they do not wound enough."

At the end of the question a member pressed the issue of whether the abandoned bullets would be considered dumdum bullets. The official being questioned said they would not be but he was contradicted by another committee member who said yes they were, the Mark IV and V split-nosed bullets were dumdum bullets.

This parliamentary exchange between legislators and government officials places the bullet in the gun chamber and not in the victim. All parties need to know that the bullet is to be loaded and fired. The global name dumdum is the result of confusion and unconcern about what the bullet does to the target beyond identifying him as the target of a dumdum bullet. Bullish dismissal of peace convention prohibitions is a sop to national interests. There must be something called a dumdum that convinces soldiers and public it is safe to load, aim and fire. The dumdum is conceptually specific and materially vague.

Equating unusually grave wounds with dumdum bullets illegally and unfairly employed when those wounds could have been caused by conventional bullets became the standard thrust and parry of warfare between any two parties with the capacity to stir the news media. The conceptual and material vagueness of the bullet made this necessary. The dumdum accusations placed a premium on examining the wounds of the soldiers and enemy ammunition and firearms with an eye to detecting the telling profile. Dumdum-like wounds in the enemy were caused by conventional bullets. Any ammunition might be modified to expand and explode upon entry.

In the conflicts where dumdums were admittedly in use there were remarkable incidents that seemed to run contrary to the established assumptions. A soldier of the 2nd Battalion West Yorkshire Regiment on active duty in northwest Pakistan in 1898 was killed by an enemy bullet fired from 1000 yards away that traveled through the folds of the great coat and the mess tin of the man in front of him before passing through his body and lodging in the cross-section of his waist belt. The recovered bullet was clearly a dumdum but had not deformed in the course of its travels. It had been captured by opposing forces during a raid on British stores.[126]

The savages could fire the conquerors' man-stopping bullets and stop the conqueror without the expanding element being activated. Dumdum bullets were infrequently recovered under the exact circumstances of their entrance into a target. Not frequently enough for faith in their efficacy to be shaken.

---

[125] Great Britain.Parliament.House of Commons (1901: 106-7)
[126] Holmes (2011:fn150)

Finding dumdum bullets among the munitions of an opponent was evidence that the opponent was violating the rules of warfare because it was important to believe the dumdum were more destructive than conventional bullets.

Japanese troops in 1904 uncovered "two kinds of Dum Dums" made for the 1891 Russian rifle (but no rifles) when they inventoried materiel seized at Liaoyang, a Chinese city looted by both sides in the Russo-Japanese war.[127] On July 5, 1905 General Linievitch sent a telegram to the Emperor of Japan informing him that during the battle of Savantse "many Russians were wounded in such a manner as to prove that the Japanese were using Dum-Dum bullets."[128] Of course the contents of the telegram were not intended to be confidential.

Japan, China and Siam had sent delegations to the 1899 Peace Conference. Tadaku Hayashi, the Japanese first delegate, observed a lack of proportion in the prohibitions enacted. A battleship could be sunk with the whole crew suffocating, he wrote in his memoirs, but the launching of projectiles with asphyxiating gases was banned.[129] The Japanese, sensing that the Russians were asserting a right to treat them as savages and fair game under conference rules, publicized the discovery of dumdum ammunition at Russian sites together with the details of the lacerations suffered by their injured troops.[130] Russian claims of dumdum wounds from combat with the Japanese were a counterclaim to energetic Japanese assertions of dumdum use by the Russians. The Japanese were becoming full-fledged members of the dumdum club.

Japanese troops use dumdum bullets to subdue the population of Australia and establish a province of the worldwide Japanese Empire in the 1904 novel, *The Coloured Conquest.*[131] Twenty-eight years later Chinese assertions that the invading Japanese were shooting soldiers and civilians in Shanghai with dumdum bullets looked back to the 1904-05 allegations against the Japanese.[132] The dumdum complex was more reputation than fact.

Another ancient empire which had military successes against Western colonial powers, Abyssinia or Ethiopia, also became enmeshed in dumdum charges. Prior Ethiopian polities had fended off Turkish and Egyptian forces with strong complements of European and American advisors, and had allied with the British against the Mahdists in the Sudan. Menelik II, pursuing an expansionist policy from his small central kingdom, clashed repeatedly with Italian colonialists trying to establish their own sphere of influence on the Horn of Africa. The battle of Adwa (or Adowa) on March 1, 1896 was a decisive setback for a reemerging Italy.

An enthusiast for the dumdum bullet in hunting big game and stopping savages ("who will generally carry more lead than a Christian"), Augustus Blandy Wylde, a British official, wrote an analysis of the battle of Adowa without suggesting that either the Italians or the Abyssinians fired those bullets.[133] He does inventory

---

[127] Two Kinds of Dum Dums, *The San Francisco Call*, September 13, 1904: 3
[128] Dum-Dum Bullets, *Los Angeles Herald,* July 7, 1905: 3
[129] Quoted by Eyffinger (1999: 161)
[130] Takahashi (1908: 179-88)
[131] Rata, pseudonym of Thomas Roydhouse (1904). Rata blames Australian women seduced by Japanese sailors and a lack of naval preparedness for the defeat.
[132] Kuei (1932: 22)
[133] Wylde (1901)

the variety of guns acquired by the Abyssinians, which must have been preserved together with ammunition to be brought back into service by the Ethiopian state ruled by Haile Selassie when it was again invaded by Italian troops in 1935.

Benito Mussolini accused the Ethiopians of firing dumdum bullets against Italian troops, and made that violation of international agreements a justification for launching poison ("incapacitating") gas canisters against Ethiopian troops. The cover of an Italian journal *Il Travaso delle Idee* for December 1, 1935 represented Haile Selassie seated at a table handing a check to a towering John Bull, the room around them filled with crates exposing the rifles they contain and small ammunition boxes marked "dumdum."[134] The British, the inventors of dumdums, could be counted on to supply the Ethiopians with the illegal ammunition. A contemporary historian concluded that the expanding bullet wounds of the Italian troops were caused by higher caliber bullets fired from "antiquated weapons."[135] Sympathy, for the moment resided with the Italians, who had avenged themselves for their loss at Adwa despite being penetrated by the forbidden bullets.

It was one feature of the dumdum complex that use of the bullets could be displaced onto people who had earlier been nominated to be their victims. The Boer and the Ethiopian instances defied the neat division into civilized armies and savage hordes that underwrote the invention and circumscription of the bullets. The American Indians, the Maori and the Hawaiians ceased to be opponent groups as they were brought into the dumdum exchange. Many other cases of involvement with dumdums by emergent and resurgent groups filled the period between the Peace Conferences and the World War I, itself an example of the extension of European categories of mayhem to the entire world.

The first atrocities of that war were accompanied by reciprocal accusations of dumdum bullets found, captured and evidenced in wounds. Reflecting on the round of such accusations that had already been made in the wars since the Peace Conferences, one writer stated a widespread conclusion:[136]

> It is scarcely an exaggeration to say that the filing of the dumdum charges has come to be regarded as a solemn rite without which no war can be regarded as properly launched.

Dumdum charges were part of the formal framework of war against which all parties strained. This writer, also like others, returned to the ability of conventional bullets to cause the wounds expected of expanding and explosive bullets, which continued to be merged into the dumdum category.

A 1915 study of the wounding effects of English rifle bullets had arrived at the finding that only the presence of a dumdum bullet in the wound could confirm that it and no other type of bullet had been fired.[137] Photographic and X-ray atlases of bone fractures caused by various bullets published at the beginning of the

---

[134] Toto Si Sconta (Everything's Accounted For) front page cartoon. Issue dated 1 Diciembre Anno XIV, 1 December Year 14 (of the new Roman Empire, a calendar initiated when Mussolini became prime minister in 1921)
[135] Nicolle (1937: 22)
[136] Carter (1914: 66)
[137] Stangardt and Kirschener (1915)

war underlined this indeterminacy by trying to formulate a bullet typology of wounds. Conditions contributing to the nature of the wounds were such factors as the nature of the propellant, the actual weight of the bullet, the area of the body struck and the protective covering. But a pre-war development in the design of bullets themselves rendered the dumdum question technically (but not politically) moot in the European conflict and thereafter.

In 1905 the German army had adopted a new service bullet, the Spitzgeschoss, or simply the spitzer from the chief feature of its design, a point at the nose. The bullet was fully jacketed over a lead core in different weights, ranging from 153 grains upward, in a .30 caliber cartridge. It shape was the result of years of research and experiment by all the bullet-using powers to find the optimum design for high velocity projectiles impelled by smokeless powder. Recognizing the value of aerodynamic engineering of larger objects at lower speeds led in these same years to the construction of engine-powered heavier than air craft. The spitzer bullets attained higher speeds, flatter trajectories and greater kinetic energy at impact than round nose bullets of the same caliber and weight.

28. Spitzer bullet half in cartridge half in cross-section. Kephart (1911: 187)

Recognizing that "a little thing" could alter any straight jacket bullet into a "jagged, lacerating mass of lead and jacket," Edward C. Crossman, a former artillery captain turned ballistics expert, commented on the tendency of spitzer bullets in flight to be deflected and "pass through tissue sideways, inflicting needlessly severe wounds."[138] As the bullet slowed upon striking the skin its heavier cartridge base exceeded the nose in momentum and tumbled forward bringing it perpendicular to the line of flight as it entered the body.

Crossman cited the big game hunting experience in 1909 of former President Theodore Roosevelt. On an expedition to East Africa Roosevelt found that a 150-grain spitzer American service bullet outperformed a 220-grain soft-point dumdum bullet when shot from the same Springfield rifle. This supported Crossman's contention that the spitzer out dumdums the expanding bullet. Trials by another canny sportsman compared the 150-grain spitzer with heavier spitzer bullets and found that the 150-grain type was "altogether too freakish in its diving and slashing in tissue."

These trials predicted how the spitzer-based service bullets adopted by major powers would act shot from repeat action firearms into flesh whether animal or human at certain distances. The spitzer was "a latent dumdum" in the chambers of guns as the world war approached.

---

[138] Crossman (1916a: 131)

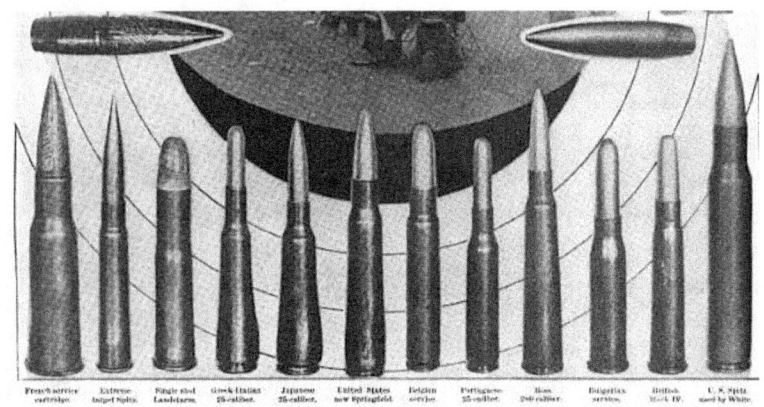

29. *The Bullets of the Fighting Nations.* Scientific American *May 1, 1915: 400*

This collection of wartime bullets arrayed against each other profiles the influence of the spitzer on the shape of the nose. The second bullet from the right, the rounded nose Mark IV, had replaced the soft-nosed Mark II and III not countenanced after the Peace Convention and by the time hostilities began had been replaced by the Mark VII. Adopted by the British army in 1911, it was a .303 cartridge 174 grains in weight with a fully jacketed spitzer-shaped nose.

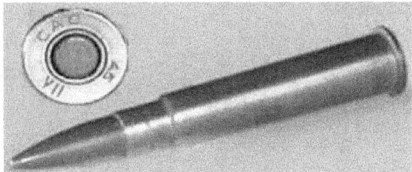

30. Mark VII *cartridge and bullet. Public domain photograph*

It didn't fall within the definition of dumdum bullets but was yet more savage in body tissues and not only because of the sharp point. The forward portion of the bullet within the jacket, instead of the usual lead, was composed of aluminum or a similarly lightweight material such as a cellulose composite, wood pulp or compressed paper. The difference in weight between front and rear, as with the spitzer, caused yawing of the front end moving into a gouging motion as the bullet entered flesh. The bullet might disintegrate under this pressure but, rest assured, the organic contents had been autoclaved to reduce the likelihood of infection.

The Mark VII remained the British service bullet through two world wars until the 1950's. It also out dumdumed the dumdums but was legal for internecine combat under the rules agreed by the combatants, all of whom had armed their troops with similar projectiles. The dumdum charges were disingenuous for those in the know. The discovery of dumdum wounds justified the accusation of dumdum bullets, and the discovery of dumdum bullets in any form suggested that the cavitated wounds and disintegrated bones were caused by dumdum bullets, which also were, of course, explosive. The U.S. Surgeon General's office headed a section of its published *Index-Catalogue* (1916): Wounds, Gunshot by Explosive [Dum-Dum] Projectiles.[139] All the worst wounds of the war caused by spitzers and Mark VII's and their ilk could be attributed to explosive dumdums made by the other side.

The combatants were also free to fabricate a range of what Crossman called "dumdum yarns" to further cast the other party in the role of hypocritical inhuman villain. There was, for instance the story, retailed in German newspapers, of the lever system attached to the side of the rifles made for the Allies, which allowed the soldier to clip off the jacketed tip of a pointed conventional bullet making it into a dumdum with lead core exposed.[140]

Lee action rifles were constructed with a magazine cut-off, a flap of steel to push in and clear the chamber. The recovery of intact .303 rounds with bent or broken tips could be taken as confirmation that this mechanized dumdum improvisation was at work. There was a crucial difference between a jacketed flat nose bullet and one from which the jacket had been removed.

A parallel accusation leveled against the German soldiers was their use of a ring in the stock of their Mauser rifles to extract the bullet from the cartridge, and reseat it in the cartridge flat end forward.[141] The field troops followed this procedure if they didn't have time to cross-cut the nose of the bullet with the service knife.

A cartoon printed in German newspapers shows diabolical figures in English and French army uniforms bent over a machine drill perforating the tips of quantities of bullets to make them explosive as a rifle-holding Death figure reaches for their production.

---

[139] The first reference after this heading was a German-language journal article on the explosive effect of low caliber bullets, not confined to dumdums.

[140] Crossman (1916a: 134). Also visible in a postcard printed in Germany showing a soldier clipping the bullet and displaying the result.

[141] Marre (1916: 24-25)

31. Death and the dumdum makers. Fn140

This cartoon was in turn printed in American newspapers along with a photograph of French dumdum bullets that had been exhibited in German shop windows to illustrate invidious German propaganda.[142]

Dumdums were made by men to be used against hypermasculine adversaries and did not involve women except as incidental victims. A cartoon published in a German satirical journal at the outset of the war took a look at English domestic life at the beginning of the Christmas season and found a group of women, girls, mothers and elders, seated around a table hearthside busily crafting dumdums from service bullets.[143] The English field dumdum conversion at Damarli was remade as an indictment of English society.

For every charge there was both a denial and a countercharge from official sources printed in the newspapers. Boxes of dumdum bullets were discovered in a captured store house; a local boy was wounded by what must have been a dumdum bullet. The other side's accusations are absurd; they are doing what they say we are doing.

This hall of mirrors criss-crossed by sniper fire was made possible by the fact that all bullets potentially could act as dumdum bullets were supposed to. Writers on one side, while seeming to take this view, could also try to exempt their side's bullets from the dumdum assertion, as Crossman did with the Mark VII and German writers did with the spitzer. No one stood entirely outside the fray and pointed to the dangers of all bullets. The result was a world-weary air of recounting the latest accusation and reviewing the history of charges extending to the beginning of the century. In 1991 a memorandum of law by a military attorney

---

[142] Swiss Super William Tell Refutes German Story of French Dumdums, *The Sun* (New York City), December 13, 1914
[143] Ein Blick aus dem Englischen Familienleben, *Simplicissimus*, December 1, 1941. See Chapter 17

evaluating yet another bullet variant of the dumdum pattern could look back over the entire history in a paragraph:[144]

> Wound ballistic research over the past fifteen years has determined that the
> prohibition contained in the 1899 Hague Declaration is of minimal to no value,
> inasmuch as virtually all jacketed military bullets employed since 1899 with pointed
> ogival spitzer tip shape have a tendency to fragment on impact with soft tissue,
> harder organs, bone or the clothing and equipment worn by the individual soldier.

The Hague Declaration didn't prevent anyone from using expanding-fragmenting-explosive bullets. It just provided cover and a language for the war of words.

The interlacing declarations and dodges of shape changing bullets are manifest in a set of statements, "called confessions by Germans" of British Army officers captured in France. Col. F.H. Neish stated he was issued "soft-nosed" ammunition for target practice, but when he was captured had only three pointed bullets borrowed from other officers, having buried the soft nose bullets. Another officer, Col. W.B. Gordon, a copy of whose statement was reproduced in the newspaper, received "flat-nosed" ammunition like that given to the soldiers for practice before deployment to France. He put his revolver in his luggage and buried the bullets. The article was accompanied by Gordon's drawing of a flat nose bullet and a photograph of flat nose bullets.

---

[144] Parks (1991: 88)

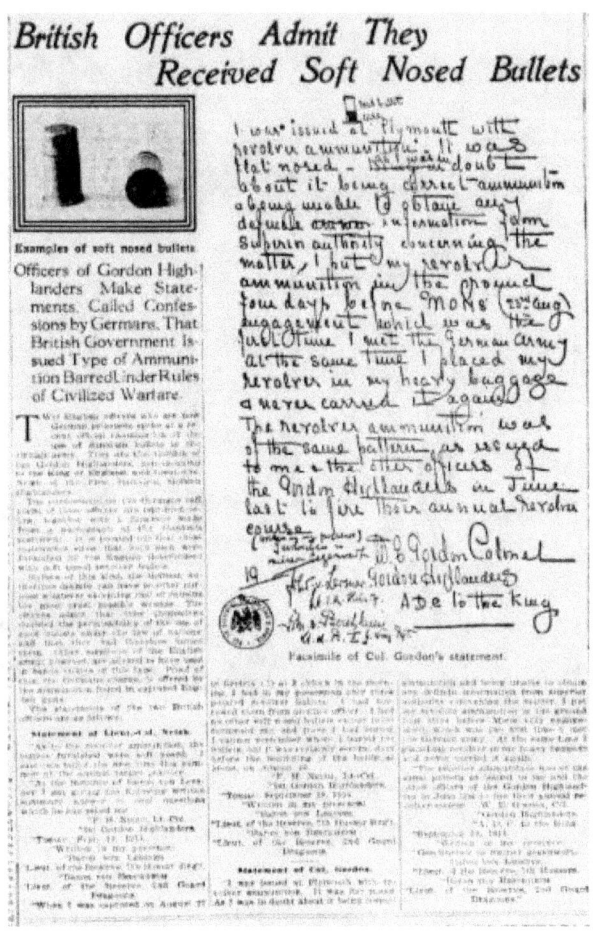

32. The New York Sun, November 2, 1914: 8

The Germans gained a propaganda benefit by getting the British officers to admit in writing they had been issued soft nose/flat nose bullets. The British displaced the stigma of being part of an army that carried the bullets by associating them with training and by saying they buried them without ever having fired them. All the Germans had to show was Gordon's little drawing. The Germans could not be accused of extracting the confessions and falsely displaying bullets the British didn't have in their possession at the time of capture. The British had never fired the illegal ammunition at their opponents. The episode is redolent of the courtesy officers of opposing sides showed each other when they could while keeping themselves on good terms with their own superiors.

The bullets kept changing shape, and exploding.

The DUMDUM Bullet as SEEN by the SCIENTIST

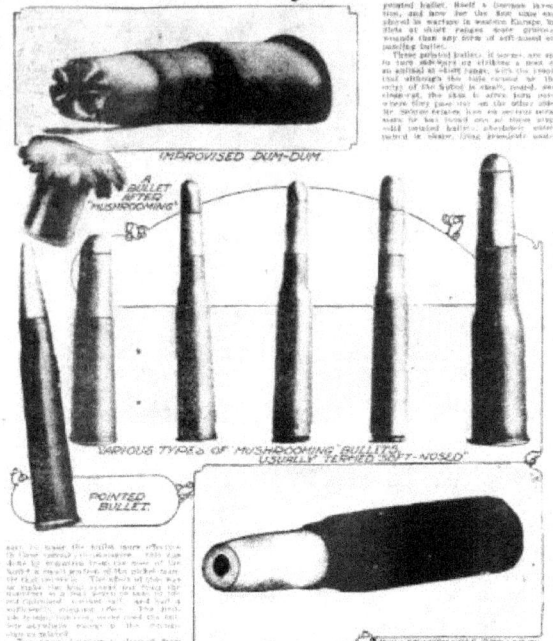

IMPROVISED DUM-DUM

A BULLET AFTER "MUSHROOMING"

VARIOUS TYPES OF MUSHROOMING BULLETS USUALLY TERMED "SOFT-NOSED"

POINTED BULLET

BULLET WITH HOLE DRILLED IN IT TO PRODUCE MUSHROOMING

The Washington Herald 12/6/1914:6

8. Humane Bullets

The French army's adoption of the 8mm Lebel rifle in 1886 marked the beginning of a trend toward a general reduction of firearms calibers by most national armies, from the existing 10.75mm to 12mm range downward to the 6mm of American naval rifles. The motive for this reduction was given as "the increased ballistic properties and convenience of the new charge."[145] The lighter projectiles carried a greater energy over a longer range of fire. As field artillery became more powerful over distances the battles were joined from afar, the focused destruction by the artillery was pitted against the finer and more widespread assault by the rifles.

It also was not coincidental to the adoption of these bullets that they weighed less, could be carried in greater quantity and would constitute less of a burden for a soldier who had other weights to bear.

The smaller caliber projectiles were not adopted because of their effects, their wounding and stopping capacity, which were largely unknown. Only combat could give a true picture, and combat after the Franco-Prussian war was dispersed and colonial, not likely to yield many cases for surgical inspection. The English expedition to Chitral, where .303 Lee-Enfield discharges did not stop the ghazis some of whom were armed with old .45 caliber rifles that did stop any English soldiers they wounded, was the occasion for the invention of the dumdum, a .303 bullet set to expand to a higher caliber on contact. This was an implicit admission of the smaller caliber's liabilities.

Beyond the rather discouraging results the new firearms gave in the field, it was difficult to gain knowledge of the effects of the bullets from tests on animals, not constructed as humans are, or from firing into corpses, since the flesh of the dead was not animated by vulnerable blood vessels like the flesh of the living and presented no opportunity to test relative stopping power.

Summarizing in 1900 the conclusions of experiments and field analyses that had accumulated over the decades since the smaller calibers became standard, August Schachner wrote:[146]

> The modern small-bore projectile is capable of producing wounds of both a humane
> and a gruesome nature.

Schachner added that the small-bore projectiles had "less disabling capacity" than the large-bore projectiles that had been taken out of service. More study was necessary to learn the capacity of these projectiles and to what extent the anecdotal evidence gathered was applicable to full-scale warfare.

The "humane and gruesome" contrastive pair had been in play for years, but it came to a point as the wars of century's end provided the testing ground for these bullets. One factor in this test was the degree to which they could be considered humane.

During the final years of the 19th century there was discussion of what would constitute "humane" warfare. This discussion was a consequence of the observed destructiveness of new weapons both those able to kill large numbers of people at the same time and those able to cause great pain to individual victims.

---

[145] Davis (1897: 36)
[146] Schachner (1900: 86)

One construction of the humane weaponry was to take enemy combatants out of the battle without causing them undue suffering or a long convalescence. The obligation of the victor to care for wounded opponents also entered into these thoughts. The question of the humane bullets was a calculated and self-serving attempt to define humaneness as the least costly means of putting an army out of commission without incurring the material and possibly moral expense of mutilated corpses. Each side adopting small caliber bullets served itself and others doing the same.

The Japanese invasion of the Chinese province of Manchuria in 1894 was promoted as a conflict between a decadent Oriental empire and a rapidly modernizing Asian state. The inefficient, corrupt and backward Chinese imperial forces rapidly crumbled before the Japanese juggernaut. Western journalists rhapsodized about the care and restraint of the Japanese forces as they advanced under the direct command of the Emperor. "Japan Finds Humanity More Effective Than Bullets In Subduing Manchuria" read the headline of one American correspondent, echoed in print nationwide.[147]

Bullets mattered too. The Manchuria war was an opportunity to assess the wounding effects of bullet caliber size difference in the field.

The Japanese surgeon Tsunesaburo Kikuchi was among military medical officers conducting tests with small-bore projectiles during the final decades of the 19th century. Kikuchi had for his consideration a rifle design developed from the Mauser by Col. Arisaka Nariakira. The bolt action Arisaka chambered a 6.5mm cartridge, a smaller caliber than the 11mm Murata rifle that had been adopted by the Japanese armed services in 1880. The Arisaka bullets were charged with smokeless powder, giving them a higher muzzle velocity than the black powder Murata bullets. An account of Kikuchi's tests was published in German in 1890.[148]

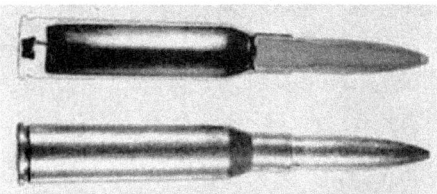

33. Arisaka cartridge cross section and profile.

In 1897 the Japanese armed services adopted the Arisaka as their main combat rifle, giving Kikuchi the opportunity to further his studies during the Russo-Japanese war.[149]

> The effect upon our enemy of our new small-bore bullet with its high velocity was, as I had predicted from experiments made previous to the war, even in the case of the slighter wounds, to put him *hors de combat* for the time. On the other hand, the extraordinarily rapid way in which the wounds heal is again to the cause of humanity. The chief cause of this is the great rapidity of the flight of the bullet, penetrating everything easily and smoothly, and causing in its passage no tearing of the parts.

---

[147] James Creelman, *New York World* December 16, 1894
[148] *Unterschungen ueber die Physikalische Wirkung der Kleingewehr-Projektile (Researches on the Physical Action of Small-bore Projectiles)*
[149] English translation in Körtner (1907: 428)

Russians wounded during the battle of Yalu were taken out of combat with a single Arisaka shot, and recovered rapidly. Chinese adversaries wounded by Murata bullets during the 1894-95 Japanese invasion suffered infected wounds. Having had the opportunity to examine many of those wounded by the Murata and by the Arisaka bullet, Kikiuchi concluded that the Arisaka quickly severed blood vessels that the slower, more massive Murata simply shoved aside, on its way to shredding organs and shattering bone, injuries requiring a much longer recovery time if recovery was possible.

Kikuchi's concept of critical wound ballistics resembled the fluid kinetics proposed by some European experimenters. While the Europeans saw the force of the bullet causing a pressure wave that did the damage inside the body, Kikuchi emphasized the effects of the release of blood from severed vessels. He viewed the hemorrhage as a powering down of the entire body through a loss of *ki* (氣). This vital essence travels through all body channels maintaining the warmth and movement of life. Once the channels are cut the body slumps into lifelessness, but it can recover with the restoration of the ki over time. One Russian, Kikuchi recounts, was shot through the lungs and lost ¾-1 liter of blood but was discharged before long.

Russians who had the misfortune of being shot a number of times were taken out of combat with the first shot, further proof of the bullet's unique disabling power. Tracks that would have proven fatal if the Murata bullet were the projectile healed much more readily when formed by the path of the Arisaka bullet.

Kikuchi's positive reviews of the rifle and bullet's performance during the Russo-Japanese war were one factor in their permanent adoption, and the continuation of variants of the Murata in reduced caliber form. The casualty record, recoverable injuries rather than deaths, enhanced the Japanese reputation as a civilized and humane fighting force worthy of admission to the ranks of those subscribing to The Hague Convention. Humane bullets allowed them to fend off the dumdum accusations that invariably accompanied that status.

The Japanese reputation for civilized conduct of warfare was perpetuated and rendered more specific during the Russo-Japanese War when Russian soldiers and medical personnel were cited by American newspapers and journals in their experience of humane bullets.[150]

A paragraph in an American periodical brought together the benefits accruing from "The Humane Bullets of Japan"[151]:

> According to a Russian medical investigator, the Japanese are using the most
> harmless bullet that ever was fired from a rifle-comparatively harmless, that is,
> in its after effects. Instead of using dumdum bullets of the deadly type surreptitiously
> brought into use in the Boer War, or of resorting to poisoned bullets, the Japanese have
> provided themselves with rifles the bore of which is so small and the velocity of the bullets
> is so great that the bullet in its flight generates heat, which enables it to act as a germicide.
> The effect of the fire is to produce anesthesia rather than a painful, lingering death. The
> bullet used by the Japanese makes a scarcely noticeable penetration in the tissue without
> tearing, and a little red spot as of an insect bite tells the cause of coma of the wounded.

From Kikuchi's point of view the bullet may have been acupunctural, but refracted into American sensibilities the Arisaka shot was like a vaccination, albeit one that caused a coma. The undisturbed

---

[150] Japanese Bullet is Humane, *Minneapolis Journal* May 19, 1904: 2
[151] *The Search-Light* July 8, 1905, 26,1:9

appearance of the skin forestalled wrenching images of the bloody battlefield lingering from previous wars. This bullet was humane because it was clean.

For those determined to hold Japan a model nation united by high values and patriotism, the humane bullet represented the spirit of the people. The journalist Alfred Stead, urging a sluggish and divided Britain on to a greater national efficiency of its own, called the bullets that disabled rather than killed Japan's way of waging war, if they must, as humanely as possible.[152] Stead backed off slightly in conceding that humanity may not have been the reason for choosing the bullets. That is no reason to assume that "humane results" were not weighed in the choice.

Rather than a less brutal conduct of warfare, the humane bullets of the Japanese might make the absurdity of warfare itself apparent to the opposing parties when they recognize that bullets can be humane. [153] Two opponents reduced to battering each other without intent of killing might well submit their differences to a court of arbitration without the "disagreeable accompaniments" of warfare. The Japanese might be precipitating an international drive to peaceful resolution of conflicts with their ammunition choice.

These were the opinions of commentators at some distance from the fighting they hoped the bullets would solve. The question of the humaneness of small-caliber bullets was alive for American military observers attached to the armies (Russian and Japanese) in Manchuria during the war that gave Kikuchi his evidence. Their comparison of Russian with Japanese bullets in terms of their humanity is laced with an irony not apparent in Kikuchi's reports.

Is it possible that the 7.60mm Russian bullet, heavier and with a slower initial velocity than the 6.50mm Japanese bullet is more humane? Affirmative, if "humane" means that the bullet is more likely to kill its target. It appears at first that the Japanese bullet had trumped the Russian in the view of the American observers.

| | Rifle. | Model (year) | Cal. mm. | PROJECTILE. | | Initial Velocity. m. |
				Description.	Weight. g.	
Japan ...	Arisaka ..	'97	6.5	Hardened lead core with copper-nickel jacket.	10.5	715
	Murata...	'94	8.0		15.42	564
Russia ..	————	'91	7.62		13.7	615

34. Comparison of Arisaka, Murata and Russian bullets. Balck (1915: 127)

In seeming confirmation of this conclusion, the observers refer to the report of Dr. Wreden, an American surgeon with the observer mission, who acknowledges that "the Japanese bullet is more modern

[152] Stead (1906: 287-88)
[153] Humane Bullets, *The Leader* (Guthrie, Oklahoma), June 1, 1904: 1. Leslie G. Niblack

and humane. About 52 percent of those wounded returned to the fighting line inside of one month."[154] Dr. Wreden found, however, that wounds within a 200 meter range of the firing were likely to be fatal because of the explosive effects of the speeding bullet. Between 400 and 800 meters wounds were less deadly except for abdominal wounds, where there was a danger of sepsis.

Taking this into consideration, the surgeon instructs soldiers that in action against the Japanese they fight only during the warm season, on soft dry ground, at distances greater than 250 meters, with empty bowels and bladder (no chance of sepsis) and with no shooting at heads. Then, he finishes, the bullets "so deadly today might prove "humane."" A rhetorical litotes. The Japanese bullet in the end, like the Russian bullet at the beginning, is humane only in its ability to kill quickly. Any reduction in casualties have come about due to improved medical field services.

Another American observer, the Navy surgeon Raymond Spear, accompanied Russian forces and gave an account of the medical and sanitary features of the campaign. The Russian doctors, often with training in German and French medical institutions, were called up from private practices at the outset of the war. The soldiers often were peasants who arrived with a religious faith they believed would preserve them from the bullets, and came away wishing that they had more education and less religion. An officer who dropped his Browning pistol and accidentally shot himself in the head recovered well after surgery with the only change in behavior being that he took up the writing of poetry.

Spear compared the Russian with the Japanese ammunition: the light, fast Arisaka ("Meidji") bullets passed through the body like long needles. Close fighting did not produce wounds as severe as were expected. In wounds of the abdomen, as elsewhere in the body, the Japanese bullets "surprised the medical men by their humanity." [155] Making several of the same observations as Wreden did on the conditions of the bullets' effects (less likelihood of infection if the stomach was empty), Spear did not give the humane repute an ironic twist like his colleague on the other side of the lines. He reflected the opinions of the Russian doctors he worked among, which were carried into American and European newspapers.

They were the Russian doctors' opinions expressed to professional colleagues, not the opinions of the officers or enlisted men. Those opinions of the Japanese humane treatment of adversaries were succinctly expressed by the instances of Russian officers attacking and killing Japanese medical staff attending to their wounds.[156]

A more detailed expression of non-medical reactions to the humane bullets came in the form of officers' and officials' ruminations on the Russian defeat in the war published in a Russian newspaper, *The Russian Gazette (Russkaya Gazeta,* 1904-6*).*[157] Captain W.H. Bingham, the translator into English, comments at the beginning of the text that censorship in Russia would normally have suppressed these opinions but the public demand for an explanation of the military failure called for some amount of plain truth.

The Japanese rifle and bullets with their light weight, high velocity, greater accuracy and flat trajectory, were superior to the heavier, slower Russian issue. Japanese troops did not have to be intoxicated with

---

[154] U.S. War Department. Office of the Chief of Staff (1906: 197)
[155] Spear (1906: 67)
[156] Edgerton (1997: 319)
[157] Bingham, trans, (1905)

alcoholic beverages to achieve their gains: their firepower was sufficient. The Japanese were able to maintain a close perimeter of fire while the Russian bullets whistled over their heads. That made it difficult for infantry to approach positions, including towns, occupied by Japanese troops. Brick walls that sheltered the Japanese from the Russian bullets did not shelter the Russians from the more deeply penetrating Japanese bullets. This having been conceded...

> It is worth while clearing up the question as regards observations on the "humane"
> character of Japanese bullets, producing as a rule slight wounds. These observations,
> in a measure, are not reconcilable with the above referred to greater penetration of
> Japanese bullets. As a matter of fact, we here meet the illusion, depending on the fact,
> that in considerations resting on the trivial nature of wounds from Japanese bullets,
> the percentage of these "humane" bullets is overlooked...To put it shortly the "humaneness"
> of the Japanese bullets consists mainly in the fact that instead of severe wounds,
> they kill on the spot, saving their victims from long agony.

There were a great many more killed than severely wounded, and many more wounded at a distance. In that sense the Japanese bullets were humane.

Military surgeons were the voice pronouncing Japanese bullets humane. They saw minor wounds and rapid recuperation; they did not count the dead. Tacticians and commanders, who had to account for troops lost, were not entirely won over by the recovery of those wounded when there was such a greater proportion of dead. For those sufficiently distant from the arena of combat, the Arisaka bullet was an opportunity to evaluate small-caliber projectile performance on the field of battle.

An identification of bullet type and intent with national character wrangled with the potential for any humanity at all in warfare. A French military surgeon, G.-S.-F. Salle, recorded his observations in a précis of a regimental conference with a title edged in irony "Humane Bullets and Their Wounds."[158]

> This easily noticed mechanical effect [stopping power of an exposed nose] had already
> occurred in 1889 to General Tweedy, the eminently British idea of how humane it is
> to have the men remove the front part of the jacket from the bullet in its cartridge.
> It was only a question, it is true, of using it against indigenous Africans, followers of the
> Madhi or others, and one well knows the minimal value the English attach to human life,
> when it is a matter of poor natives who have the misfortune of finding themselves on the
> path of the English civilizing mission.

This impressionistic sense of another nation's sense of its own humanity at the soft point of a bullet was one more swing of opinion among professional military men. The ammunition for the coming larger wars was being prepared.

William Balck, a German army officer whose multi-volume tactical treatise went through four editions between 1904 and 1915, epitomized the division of opinion between military surgeons and tacticians when he referred to the ballistic advantages of the small-caliber projectile. These advantages, undesirable from the viewpoint of surgeons, derived from the tendency of pointed-nose, small-caliber, light-weight bullets (German S-bullets, French D-bullets, Arisaka) to tumble upon impact, causing serious wounds. "These wounds, while not inhuman, instantly disable the man struck, or, at any rate, postpone his recovery indefinitely."[159]

---

[158] Salle (1899:9)
[159] Balck (1915: 130)

Selecting the "not inhuman" over the "humane," the German general staff elected not to reduce the caliber of the standard issue bullets below 8mm.[160] At that size and with the velocity achieved with smokeless powder and with a jacket design most likely to promote tumbling the ammunition for the upcoming war satisfied The Hague Convention without being "humane." Not humane and not inhuman did not mean not dumdum.

The advent of World War I, such a lively environment for dumdum bullets, was the final stopping line for humane bullets. The phrase cropped up rarely in descriptive verbiage of the war's wounds and the ammunition that caused them. Medical staff no longer used the language of humane intentions for any type of bullet and the wounds they caused. A singular recollection of the Japanese humane bullets of the Russo-Japanese War recurred to the observers' reports that found them more lethal than the Russian bullets.[161]

Without reference to Japanese bullets or small-caliber bullets in general, G.H. Makins denounced the very idea that ammunition could be humane. [162] Makins was a battlefield surgeon, reputedly the first British surgeon to be sent to organize medical services in South Africa during the Boer War. He wrote a number of papers and a surgical memoir based on that experience.[163] With the thrust of the German army that initiated World War I in Europe he joined the British Expeditionary Force and served through the war as a medical officer. He continued to produce papers and books on case histories and surgical treatment.

Like surgeons from Paré onward, Makins sought to improvise treatments that answered the variety of injuries among soldiers and to convey his discoveries to colleagues. He trained a skeptical eye on outlandish beliefs about the war that arose among the public back home. He wrote a brief piece published in the *British Medical Journal* on "the so-called "poisoned bullet," which turned out to be nothing more than a wax coated bullets captured from the Boers.[164] The wax, transparent or green in color, served as a lubricant coating to reduce erosion of gun barrels. Makins didn't bother to deny that it was poisoned, which should have been obvious. He did not even broach the subject of dumdum bullets in his memoir or papers.

One product of his four-year sojourn in France was a treatise on gunshot injuries to the blood vessels.[165] In the first pages of the book he contrasted the vascular injuries observed during the Russo-Japanese war with those suffered on the European battlefield. The bullets fired in Manchuria caused clean perforations and limited local lesions. Three specimens of gunshot wounded blood vessels from that war he included in a photograph he read as evidence of this. Not having been present during that war he gained a surgeon's perspective by looking back to the appearance of the specimens when they were first struck by bullets.

The high-explosives and fragmented projectiles of the recent war, new rifle profiles and the discharge of machine guns brought "an augmentation of the degree and extent of contusion and an increase in the number of incised and lacerated wounds." Shrapnel from exploding shells and fragmentation of the bullets themselves due to ricochet more than in earlier wars magnified the severity and complexity of the wounds.

---

[160] Körting (1907: 406)
[161] Problems for Medical Men in Present War, *The Sun* (New York) January 3, 1915: 2
[162] Articles in *The Lancet* reflected in a newspaper article, The "Humane" Bullet: A Report That Makes Its Alleged Humaneness Mostly a Farce, *The Washington Herald*, January 10, 1915: 6
[163] Makins (1901)
[164] Makins (1900)
[165] Makins (1919)

For someone with Makins' experience, humane bullets were truly a farce. The set of circumstances that had made it possible for Kukichi to project a speedy recovery after a bullet-severed blood vessel no longer obtained for anyone.

9. Soft Nose Bullets in the Mexican War

Terminologies of bullet design, technology and wounding history are associated with each other and defined apart during historic periods. Dumdum bullets and humane bullets turn about each other and separate. At times specific bullet descriptions are used to the exclusion of other categories that might seem to an outside observer to apply to the bullets named. Soft nose or soft point bullets, those manufactured with the jacket exposing the lead core, were sometimes classified as dumdum bullets, and sometimes were a type without the dumdum label.

They also intersected with the humane bullet category, but only with reference to their use in hunting.[166] Sport hunters were warned by bullet specialists that "any soft nose bullet should have its length, strength of mantle, temper and relative exposure of the tip carefully proportioned to the power of the gun and the character of the game hunted."[167]

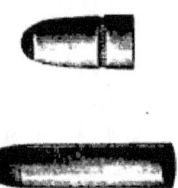

35. Short and long soft nose bullets. Kephart (1917: 485)

The long soft nose bullet is preferable to the short one because the long bullet mushrooms at its tip only while the short may mushroom along its entire length,

36. Mushroomed short soft nose. Ibid.

go to pieces and shatter bone without penetrating. The short bullet is best for soft skinned animals and not good for hunting dangerous game since it is likely to mushroom on first impact and not enter far enough to stop the animal. The small caliber short soft nose is for exterminating vermin and not for squirrels, rabbits and game birds since it is likely to wound but not kill them with the first shot, allowing them to escape and die a lingering death. Safety and care considerations help determine the subtype of soft point.

---

[166] For instance: High Pressure Gun is Humane for Big Game, *The Bemidji Daily Pioneer* (Minnesota), November 23, 1907: 1; Small Calibre Rifle Wounds, *Peninsula Enterprise* (Accomac, Virgina) February 20, 1897
[167] Kephart (1917: 485)

The bullets in their longer form were considered a humane way to put down injured livestock, horses and large zoo animals. The refinement of soft nose bullet types embraces a calculus that factors in hunting and animal husbandry practices with the weight and shape of the bullet and the kinetic energy of its flight.

No such refinements figured into the analysis of what were suspected of being soft nose bullets shot into humans. Soft nose bullets were deliberately shot into animals for humane hunting and management purposes. What could be the intent of shooting them into humans, or the reason for suspecting they were shot into humans?

The mushroomed bullets found in the body of murder victim George E. Bailey led the prosecutor in the 1901 trial of John C. Best to suggest that soft nose bullets were used. A state police investigator gave testimony defining soft nose bullets and distinguishing them from unfired bullets found in the possession of the accused. [168] The accused was convicted and executed for the murder because testimony proved he had fired a gun at the victim before his death from gunshot injuries, whatever bullets he may have used.

A member of the large posse pursuing escapees from Folsom State Prison in California was found with a gaping wound to the skull that had spread out his brains.[169] A mass of buckshot in his lower jaw did not deter the inference that the skull wound was the exit wound of a soft nose bullet fired by a convict because criminals fired soft nose bullets into people.

In a Senate hearing on the "affray in Brownsville, Texas" spurred by soldiers during two nights of August, 1906 an army officer testified that the soft nose bullets they fired off, recognized because they mushroomed, were hunting bullets, not combat issue.[170] An article on soft nose bullets in a medical journal was entirely about deer hunting.[171] Soft nose bullets were for hunting, though their path in animals might yield signs for detecting them in humans.

Enraged murderers, escaped convicts, drunken soldiers, a jealous "Negro" who unloaded the soft nose bullets in his revolver in a gun fight with police[172], these were the people likely to be thought to fire off soft nose bullets at humans rather than animals. The spent bullets found at the scene or in the body, and the wounds observed, were taken as evidence this type of culprits was responsible.

Dr. Watson examines the dead body of Ronald Adair found in an upper storey room that had been locked.[173]

> No one had heard a shot. And yet there was the dead man, and there the revolver
> bullet which had mushroomed out, as soft nosed bullets will, and so inflicted a
> wound that must had caused instantaneous death.

Sherlock Holmes, emerging from disguise, tricks Col. Sebastian Moran, the killer, into firing at a wax model of himself visible in the window. Col. Moran, a big game hunter and confederate of the now deceased

---

[168] Massachusetts.Superior Court (1901: 584)

[169] Two Convicts Open Fire on Officers at Dutch Flat, *The San Francisco Call* August 3, 1903: 1

[170] United States. Congress. Senate (1908: 2690)

[171] Crutcher (1906)

[172] Negro Woman Held in Tragic Shooting, *Evening Public Ledger* (Philadelphia, Pennsylvania) December 4, 1916: 4

[173] The Adventure of the Empty House (1903), the first Sherlock Holmes story published after the presumed death and return of the sleuth.

Moriarity, had used his silent air gun to project into Adair from the street below a single soft nose bullet, achieving a seemingly perfect kill. In this story set in 1894 Dr. Conan Doyle imagined the evidence for the transition from animal to human game in the shape of the bullet and the wound.

In the early battles of the Mexican revolution, Casas Grandes and Agua Prieta, reports from Mexican federal field commanders received in El Paso, Texas and passed on by correspondents to American newspapers, attributed the wounds received by the men under their command to soft nose bullets fired by the rebels.[174] The same newspaper articles also included word from a correspondent wired from Hermosillo, Sonora that he saw (Mexican) federal troops march out "with their belts full of soft nosed bullets." A year later the editorial page of an El Paso newspaper remarked that "hundreds of thousands of soft nosed bullets" had been imported by the rebels. [175]

A doctor who examined the dead in the aftermath of the battle between the forces of Francisco Madero and Mexican president Porfirio Diaz near the rail junction of Agua Prieta, Sonora in April 1911, made a distinction between the type of wound suffered by each set of forces. Dr. F.E. Shine told a reporter for the *Bisbee Daily Review* (Bisbee, Arizona) that the soft nose bullets used by Madero's rebels were "inhumane."[176] They could take off a limb or leave a wound cavity large enough to stick a fist in. Dr. Fine believed that the steel, fully jacketed bullets fired by the Mexican federal troops were "more humane" since they passed directly through the flesh and only caused severe injury when they struck a vital spot. Yet he granted that the soft nose bullets were the only ammunition available to the rebels.

The medical reportage from this early battle of the Mexican revolution introduced the application of two current categories to the ammunition used: soft nose bullets and humane bullets. The wounds as viewed by a doctor were the chief evidence for both. It was consistent with soft nose bullet vernacular that the undersupplied rebels, not viewed with sympathy by Americans in Arizona across the border from Agua Prieta, were the ones firing the soft nose bullets. The steel bullets of the federals, like the Arisaka bullets of the Japanese in the Russo-Japanese conflict seven years earlier, were more humane. They were not likely to be found otherwise, as the Japanese bullets were, if there was an inhumane party already in the struggle.

Both the soft nose bullets and the steel jacketed bullets were imported by the Mexican factions from American manufacturers. Diaz, a reformer before he assumed dictatorial powers, had professionalized the army. The military grade steel jacketed bullets were part of this process. When Madero became president after the flight of Diaz in June, 1911, he inherited an army that used the bullets his supporters had been dodging.

In March, 1912 with the continuation of the unsettled state in Mexico, the United States Congress passed legislation designed to prevent Americans from abetting the conflict.

A subcommittee of the Senate held hearings in September of that year in El Paso, Texas to investigate compliance with the law. One hardware dealer told the committee chairman, Senator William Alden Smith of Michigan, that his firm had sold "soft point bullets" to some of their usual customers and that he had

---

[174] Forbidden Soft-Nosed Bullets Used in Mexico, *The Logan Republican* (Utah), March 23, 1911: 3
[175] *El Paso Herald,* March 11, 1912
[176] Soft Nosed Bullets Inhumane, *Bisbee Daily Review*, April 23, 1911: 8.

learned from newspapers that Orozco, a revolutionary leader in the state of Chihuahua, was paying high prices for ammunition carried across the border.[177]

D.M. Payne, who identified himself specifically as a seller of arms and ammunition, was questioned more closely. Payne acknowledged that soft point bullets are used for hunting game, and that they mushroom when they meet resistance. The full metal jacket bullets manufactured for the military go straight through an object they strike.[178]

> Senator Smith. In either of these revolutions [Orozco's or Madero's] have the
> purchasers of ammunition asked for any special kind of bullet?
> Mr. Payne. The Mexican Government has always specified they should be the
> military cartridge or full metal patch.
> Senator Smith. How about the present revolutionists?
> Mr. Payne. They in some cases have asked for soft-point bullets and sometimes
> for the full-jacketed bullets.
> Senator Smith. Is the soft-point bullet permitted in civilized warfare?
> Mr. Payne. I understand it is not.
> Mr. Paiz. I was shot with dumdum bullets, and I know my own men used them.

This ends the line of questioning on ammunition types. In the manner of congressional hearings no accusations are directed against Payne, but the revolutionaries' inclusion of the forbidden soft point bullets in their armament is established. The matter of bullet cost, which Dr. Shine found explanatory if not exonerating, is never raised.

The subcommittee hearing did as much to make the Mexican revolutionaries out to be soft point bullet users as it did to locate their American suppliers. The source of the bullets was not named in any of the pieces that attributed dastardly soft nose bullet attacks to the rebels. In some accounts the bullet fire seemed to come from Mexico itself.

Some American government officials and business figures, property owners in Mexico facing an agrarian revolt, in early 1914 pressed President Woodrow Wilson to annex Mexico or at least declare it a protectorate. The subsequent presidential order only authorized an occupation of the port city of Vera Cruz and to enact an arms embargo, securing the armaments stored and being delivered there by a number of European and American firms.[179]

Jack London was commissioned by *Collier's Magazine* to accompany the American force that arrived to carry out the order. London's social Darwinist perspective on the Mexican peons made them out to be the "oxlike" result of generations of selection for dumb obedience to higher authorities. Mounted on the rooftops, they did shoot Americans.[180]

> Also, the mark of the cross, rightly applied to the steel-jacketed nose of
> the bullet, can turn that bullet into a dumdum that makes a small hole on
> entering a man's body and a hole the size of a soup plate on leaving. It
> requires no intelligence thus to notch a bullet. Even a peon can do it.

---

[177] United States. Congress. Senate (1913: 122-23). Adolf Krakauer testimony.
[178] United Statees. Congress. Senate (1913: 180)
[179] Hart (1987: 290-94)
[180] London (1914a: 10)

London saw the Mexicans' successful sniping as the result of high velocity bullets modified by the peasants themselves, and not manufactured soft nose bullets delivered from overseas. In another of the several articles he sent to *Collier's* on the Vera Cruz action he witnessed the amputation of the leg of a Mexican peon that had been shattered at the thigh joint by a "wobbling, high velocity American bullet" which evidently had not been made into a dumdum.[181]

An American Press Association photographer named Adrian C. Duff was on board an American battleship when it docked in the Gulf of Mexico port of Vera Cruz as American forces moved to occupy the city.

In addition to his photographs Duff sent his newspaper syndicate a set of written "eye pictures" of the American servicemen as they entered the city. "One of those treacherous Mexican soft-nosed bullets" got Chief Gunner's Mate Boswell in the stomach, a wound from which the popular crewman died soon afterward.[182] Duff made an equally graphic connection between wound and bullet type when he arrived at the Vera Cruz railroad station to find an expiring "bluejacket" (navy sailor) who had his head ripped open by one of the deadly bullets fired by a sniper.

Duff refers to the bullets as "Mexican" but since the mission was intended to prevent the landing of American-made bullets he must be naming them after the Mexican soldiers or naval cadets who were firing them. Some of the cadets at the Mexican naval academy in Vera Cruz had elected to remain and resist the American occupation, and the academy buildings were bombarded and severely damaged. Unfortunately, the federal military commander also released convicts to take part in the defense of the city. The first Mexican newspaper account of the occupation of Vera Cruz does not distinguish bullet types in its reports of shootings by the belligerents.[183] It only recounts the sudden and inexplicable deaths of Vera Cruz citizens by gunfire. Mexican histories of the revolution generally accuse North Americans of firing dumdum bullets or expansive bullets in armed encounters.[184]

In 1906 strikers at the Sonoran copper mining operation Cananea, owned by a Mexican subsidiary of the American firm Anaconda, were met by what was described as dumdum fire from Arizona Rangers and Mexican rural guard (*rurales*).[185] A peaceful protest by Mexican workers against wage discrimination-American workers were being paid twice what the Mexicans earned per hour-escalated to violence and was suppressed with gunfire. Long before the occupation of Vera Cruz both Mexicans and Americans were shooting Mexicans with dumdum bullets.

The Arizona Rangers, like the Texas Rangers and Fullerton's Rangers in New Mexico, carried a mix of soft nose and steel jacket bullets for their Winchester rifles.[186] Dave Anderson, who had been a Texas Ranger, an Arizona Ranger and a county sheriff before signing on as bodyguard to Col. W.C. Greene, the

---

[181] London (1914b: 6)
[182] Duff (1914)
[183] McNelly, trans. (1914)
[184] Sayeg Helú (1996: 138)
[185] Diaz Cárdenas (1989: 61)
[186] Hornung (2005: 80); Ivey (2010: 176); Alexander (2009: 306)

chief executive of the mining company, had a role in breaking the strike.[187] Facing a revolt by the Mexican workers, Greene telegraphed Arizona requesting armed volunteers to reinforce the rural guard, and an Arizona ranger captain crossed the border against the orders of the territorial governor at the head of a 275-man posse. The dumdum or soft nose bullets that killed some of the strikers introduced into Mexican conflicts a pattern that varied according to the biases of the observer and the bullets available.

The inhumane soft point bullets of the rebels and the steel jacket bullets of the federal troops in Dr. Shine's report on Agua Prieta in 1911 were both used by Mexican fighters perpetrating the "Santa Ysabel massacre" in June, 1916. El Paso physician F.E. Miller had an opportunity to examine several of the bodies of the 18 American mining businessmen who were killed by soldiers as their train traveled to a mine they planned to inspect with a prospect of reopening. They never reached the mine. Their train was stopped by a derailed car on the track ahead and they were forced to leave the car.

Decrying "the brutal methods of the Mexicans," Dr. Miller's report "shows the use of soft-nosed bullets in inflicting death wounds while Mauser bullets were used to bring down the victims who attempted to escape."[188] For some death by knife, bayonet, blunt instrument or bullet was followed by further damage to the body. The head of one man (each of them is named) was shattered by a soft nose bullet, beaten and then shot with more soft nose bullets after death. Fragments of lead "like bird shot" inside the chest skin of a man were the remains of a soft nose bullet fired into his head from above. A Mauser bullet in the back ended the life of a man who already had been shot twice. He was then bayoneted and shot with both Mauser and soft nose bullets after death. The division of labor between Mauser and soft nose bullets described by Dr. Miller broke down after the men were dead.

The deliberate post-mortem mutilations of the train passengers were not explained in any way other than as the result of the nature of the Mexicans. Rebels were no longer distinguished from federal troops: antagonisms were expressed against Mexicans in general. A Senate hearing that took place sometime after the massacre concluded that both rebels and soldiers were involved in bringing it about.[189] The forces of Pancho Villa were considered the most likely perpetrators. After Villa's raid on Columbus, New Mexico the previous March, a punitive expedition under the command of General John Pershing crossed the border in pursuit of the raiders and remained in Mexico for nearly a year. American infantrymen soon discovered "American-made" dumdum bullets ("soft-nosed bullets of the dumdum type") in the belts of dead Mexican bandits.[190] This should not have been very surprising since one of Villa's motives for attacking Columbus was that he had been cheated by the local hardware dealer in an arms purchase and raided the town to secure firearms and ammunition.[191] Villa's raiders already had American-made ammunition when they attacked the town, and they acquired more as a result of their raid.

---

[187] Alexander (2003)

[188] Brutal Methods of the Mexicans: How the Americans at Santa Ysabel were Put to Death, *The Ogden Standard* (Utah), January 20, 1916: 8

[189] U.S. Congress. Senate (1920: 2757-60))

[190] Says Bandits Used Dumdums, *The New York Times,* March 15, 1916; Villa Used Dum-Dums Made in U.S. in His Raid on Columbus, *The Omaha Daily Bee,* March 15, 1916: 1

[191] Primley (2003: 213)

American firms supplying soft nose dumdum bullets to Mexicans in their current turmoil came to a head with the visit of California congressional representative Julius Kahn, ranking member of the House Committee on Military Affairs, to the Mexican border region of New Mexico in late 1916. Kahn told a reporter that he discovered quantities of ammunition, including dumdum bullets, smuggled across the border. [192]

> If ever our troops are compelled to campaign there, they will be butchered by dumdum
> bullets sent to Mexico by renegade Americans out after profits.

Perhaps Kahn was not being disingenuous in neglecting to mention that the American expeditionary force under Pershing's command, armed in part with soft nose bullets, was still on the other side of the border. He soon was demanding a congressional probe of the border activity and the involvement of large American firms.[193]

Pershing's expedition did not succeed in capturing Villa, though it did disrupt supply lines and kill several commanders. A diplomatic solution was arranged and the troops were recalled in advance of the formation of American Expeditionary Forces sent to Europe beginning in May, 1917 under the general command of Pershing. As contingents of American troops parted for the war, there were clashes between Mexican and American garrisons stationed at the border. The Battle of Ambos Nogales on August 27, 1918 was precipitated by a gun runner trying to cross the border without submitting to inspection at the American customs post.

During the Senate hearings conducted in 1919 to investigate "Mexican affairs" over the preceding years the subject of ammunition rarely came up. Senator Albert Fall of New Mexico questioned a 17-year old Mexican man, Jesus Paiz, who 3 years earlier had held the reins of the horses during the Columbus incursion.[194] Paiz said he almost had his leg shot off with dumdum bullets. When the senator asked him who shot him he said it was the men he came with, not the American soldiers. "I don't think the American soldiers used dumdum bullets, did they?" Paiz asked. The senator said he never heard of it. There were no further questions on the subject.

With the end of the war in Europe and the cessation of cross-border confrontations the public accusations of soft nose bullet or dumdum use also declined. The demand that Mexico be annexed or at least be made a protectorate faded as American interests gained control of Mexican oil and other natural resources. American troops were no longer faced with the prospect of crossing the border to be met with American-made bullets.

The echoes of the former dumdum discharges were heard in several assassinations. The Villista general Felipe Ángeles died on November 20, 1919 of gunshot wounds caused by dumdum bullets.[195] Pancho Villa himself retired the following year and established a settlement for his former soldiers. He was on the way to his bank in Chihuahua on July 20, 1923, unaccompanied by his usual bodyguard when his car was fired upon

---

[192] Dumdum Bullets Sent into Mexico, *The Washington Times*, December 3, 1916: 7
[193] Demands Probe of Border Smuggling, *The Washington Times,* January 8, 1917: 6
[194] U.S. Congress. Senate (1920: 1620)
[195] Taracena (1987: 246)

by men standing in the road holding rifles. Seven dumdum bullets were found in his body.[196] Villa had announced his intention to reenter national politics with a run for the presidency.

---

[196] Katz (1998: 766)

10. Lower Nervous Organization

> He agreed that the Dum-Dum bullet should not be used against civilized
> troops, but when one came to deal with those of a lower nervous organization,
> such as savages, Soudanese, Ghazis in India, the Dum-Dum bullet was the only
> means of stopping a rush.[197]

The conference speaker immediately before Surgeon-General Harvey, who made this statement, deplored the invention of the dumdum bullet and expressed the belief that the British nation was being repaid for that invention in the current South African war. For that speaker the dumdum was a moral deficit, while for Harvey it was the only way for civilized men to overcome attackers of a lower nervous organization.

The phrase "nervous organization" was used to set out types and degrees of responsiveness to sensations in animals and humans. Fish generally had a low nervous organization; some, but not all, horses did as well. Women had a "delicate" nervous organization, which when suffering the inevitable damage could be repaired by resort to medication. The original colonies of the United States, in one formulation, had a low nervous organization, like that of a jellyfish, capable of being separated into pieces and still surviving. By the early 20th century railroads, telegraph, telephones and intricate interstate commerce had formed a true nervous organization for the nation.[198] Nervous organization was a stage of personal and societal development, a psychological continuum and a metaphor for emergent technologies more closely approximating biological systems.

Nervous organization was defined as organismic sensitivity characteristic of an integrated whole. It could be refined and not strong, as in women, or strong and not refined, as in savages.

For those who placed the contest with savages and others belonging to that category in the physical realm, the savages had all the advantage. They had, after all, achieved dominion over animals strictly through strength and use of weapons. "It is surprising," wrote Charles Morris not long after the publication of Darwin's *The Descent of Man* (1871), "what muscular power, what endurance, what bodily agility and dexterity in the use of primitive weapons, are acquired by savages, hardened by their life in the open air, and by their constant encounters with wild beasts and hostile men."[199] It is best not to take civilized man as an example and imagine his helplessness in the state of nature. Better to think of the savage as the paragon of early humanity.

When civilized men faced these savages they had to make up for the differential. Dumdum bullets, repeat action guns, high-velocity powders and the like were offered as compensation for the lack of muscular preparedness when faced with men who lived in the state of nature and whose nervous organization approximated that of animals, only with a greater element of calculation and superb handling of weapons. This contest between the muscularity of savages and the technology of the civilized was repeated as class conflict closer to home.

---

[197]Remark by Surgeon-General R. Harvey, Director-General Indian Medical Service. Some Remarks by Way of Contrast on War Surgery Old and New, *The Lancet* (August 10, 1901) 2,1: 401
[198] State and Nation, *The Commoner* (Lincoln, Nebraska), January 11, 1907: 1
[199] Morris (1873: 534)

The savages placed at the lowest cultural level or the highest physical level, whichever was emphasized, were grouped with others possessed of a similar response pattern. Decisive bodily motion was not preceded by careful consideration. It was reflexive action not policed by higher thoughts or advanced institutions. Savages could be ranked according to various contrived physical measures as constituting a distinct descriptive category approaching measures characteristic of animals. Yet the appearance of ghazis, juramentados and amoks among populations of the civilized had to be the result of the low level of nervous organization still present and capable of taking hold. Physical primitivity and nervous primitivity were elements escaping absolute definition to form the mix of modern humanity.

This fluctuating set of identifications has been formulated and reformulated to yield ideas of anatomy, physiology and neurology broadening to encompass a wide humanity and narrowing to serve dominance by a socioeconomic group. Measurements of intelligence were engineered to favor the categories of achievement of their authors. The concept of the underlying reptilian brain allowed for outbursts and perceptions of a lower nervous organization rising from beneath the more considered cerebral cortex into an excited, competitive contemporary world. A gambler who shot another gambler to death over a game displayed the low nervous organization of his calling, a brutal insensitivity, when he refused to hold the dying man's hand.[200]

British boxing writers disdained as unfair competition African-American boxers attempting to enter the ring during the early twentieth century against British pugilists. One columnist described them in words that reflected a resentful popular Darwinism:[201]

> ...round cast-iron head with no chin, ribs that almost meet across the stomach,
> the ape's curved convulsive clutch, and a lower nervous organization indifferent
> to pain.

No gentleman fighter with a respect for the rules of engagement could be expected to prevail against such a creature. Jack Johnson's defeat of Jim Jeffries on March 4, 1910 in a heavyweight title bout was deemed the result of a match between a man and a near-animal. Johnson did not feel Jeffries' well aimed blows just as savages and animals of a similarly low nervous organization did not feel bullets. The civilized were in danger of succumbing to lower types in the boxing ring and on the field of battle if they did not adopt preventive measures.

Of the .30 caliber/156 grain bullet fired by the Krag-Jorgensen rifle, one commentator explained, while "effective enough against civilized men of a highly strung nervous organization," was of little use against savages.[202] Any penetrating missile would be sufficient to cut the strings of the high strung.

Dumdum bullets tore away the advantages those of lower nervous organization had in encounters at the borders of civilization and at its center. They were intended to make up for the deficit in reflexes and strength of the civilized facing the savage. Savages in the cities called for a protective shield of dumdum bullets more urgently than the colonial battles used to justify the bullets in the first place.

---

[200] The Gamblers and Their Ways, *The New York Times,* November 27, 1904
[201] Roberts (1985: 51)
[202] The Dumdum Bullets, *The San Francisco Call,* August 2, 1899: 6

"THE GUARDS HAVE DUM-DUMS" read an 1899 newspaper headline.[203]

> CHICAGO, Aug 8. It has become known here that every guard on the
> penitentiary walls at Joliet is armed with a new Mauser rifle, and the
> rifles are loaded with dumdum bullets, the bullet denounced at the
> Peace Conference at The Hague as a barbarous implement of warfare.
> It is said to be the first instance of its use in the United States. The bullets
> used at Joliet are made by an American firm.

The Mausers, however, were made in Germany. This same model rifle supposedly loaded with dumdum bullets, was being fired by the Boers against British troops in South Africa. The bullets, at least, were made in America, as were the criminals.

Joliet Correctional Center was at the time the largest penal institution in the United States in inmate population. The potential helplessness of the guards during a large-scale riot made the use of dumdum bullets seen necessary; the reference to the barbarous nature of the bullets suggested the gravity of the situation. The sudden fear instilled by a rushing savage was already a communal dread before the fanatical foreign fighters entered the news stream. Now the more immediate danger, the criminal, was brought into the dumdum complex. The guards had to defend against a concentration of lawless men who would stop at nothing.

Criminals could appear anywhere, the result of an individual's induced moral failure. The psychologist Havelock Ellis delineated the sources of criminality in a book that went through four editions between 1890 and 1915.[204]

> Criminality, therefore, cannot be attributed indiscriminately to the lowest
> of races. It consists of a failure to live up to the standard recognized as
> binding by the community. The criminal is an individual whose organization
> makes it impossible to live in accordance with this standard, and easy to risk
> the penalties of acting anti-socially. By some accident of development, by some
> defect of heredity, or birth or training, he belongs as it were to a lower social
> state than that in which he is actually living. It thus happens that our own criminals
> frequently resemble in physical and psychical characteristics the normal individuals
> of a lower race. This is the "atavism" which has been so frequently observed in
> criminals and so much discussed.

The criminal arrived at his distinctive appearance and manner of acting as the result of hereditary or environmental influences. This appearance was identical with that of savages who had not formed a civilized community. The criminal had ceased to participate in a civilized community or never had been able to. The criminal and the savage both had the same look. This comprised the mental picture of an assailant, a cheat, a thief.

---

[203] *The Seattle Star*, August 8, 1899: 2. Other newspapers reporting the dumdums at Joliet quoted a *Chicago Tribune* article that did not mention the Hague Peace Conference.
[204] Ellis (1915: 251). Wey (1891: 277) quotes this passage in the criminal anthropology section of the U.S. National Prison Association 1890 Meeting Proceedings.

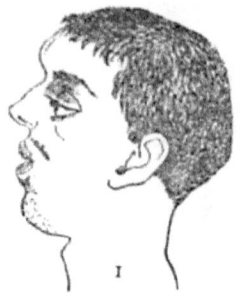 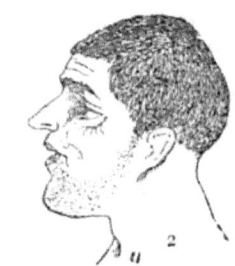

36. Line drawings of criminals from photographs. Ellis (1892: Plate II)

The criminal physiognomy was atavistic, corresponding to the face line of the savage and the ape. A Russian student of criminal physiology working in Paris invented a physical measure that yielded a gradation from refined humans to animals.[205] The cranial-mandibular index was the ratio of the weight of the mandible to the weight of the cranium, in effect quantifying the portion of the cranium occupied by the jaw. Orchansky tabulated the index values from Parisians to apes, equating French assassins with Negroes followed by microcephalic idiots on the way to apes. The index proposed quantifying the ranks of humanity this time according to visible jaw mass and the degree of animal-like muscularity of bite. This information can only have been gathered by separately weighing the mandibles and crania of identified individuals.

Parisians—women,	.	.	.	.	12.8
Parisians—men,	.	.	.	.	13.4
French assassins,	.	.	.	.	14.78
Negroes,	.	.	.	.	15.69
Microcephali,	.	.	.	.	22 to 25
Anthropoid apes,	.	.	.	.	40 to 46

37. Cranial-mandibular index for select groups. Fletcher (1891: 23)

There was disagreement about how deep-seated and inalterable the atavism was. A loud voice among criminal anthropologists, also heard in the public media, proclaimed the only solution to the threat posed by criminals was to kill them.[206] By execution if found guilty of a capital crime, or quickly and decisively with firearms if rioting in a prison.

Not many newspapers carried an article in which a secretary to the governor of Illinois, where Joliet is located, denied that the guards at the correctional facility were armed with dumdums, and dismissed the entire notion as "ridiculous."[207] The dumdum bullet complex rapidly expanded to include ape-jawed criminals among the potential targets.

---

[205] J. Orchansky as related by Fletcher (1891: 23)
[206] For example, Mental Malformation, *The Roanoke Times*, June 18, 1892: 11,"Many Criminals are Morally Irresponsible, but Society Must Protect Itself by Killing Them if Necessary."
[207] Those Dum Dum Bullets, *Daily Public Ledger* [Maysville, Kentucky], August 1, 1899: 4

It was difficult to learn what prison guards were armed with: their weapons were not available for inspection outside the prison, and the bodies of dead inmates were not examined for signs of expanding bullet wounds. When a riot at Joliet was given news coverage decades later, it was Illinois state troops who opened fire on the rioters, killing one.[208] Seeing a number of women about the men's prison, the warden had halted visits by outsiders, precipitating a violent protest. The discovery of boxes that had contained "the exclusive Silvertip bullet with controlled expansion soft-jacketed tip" was the only evidence critics of the response to the 1971 Attica, New York prison uprising could discover to support the charge that dumdum bullets had been used.[209] Bullet names had changed to make the dumdum label seem archaic.

The ammunition issued to prison guards was a sensitive point in the societal selection of criminals as dumdum-worthy because of the concentration of criminals in prisons, and their ability to engage in collective action. That simply enhanced their inclusion with savages and antagonistic warriors in the class of those of a lower nervous organization. The mentally ill, "idiots," and members of ethnic and racial minorities were also placed in the lower nervous organization category. The inclusion of laborers on strike among those deserving the bullets raises the suspicion that they are being treated as a socioeconomic class. That class was a neural class rather than entirely a socioeconomic one because individuals of any class might descend to a state of organization leading to a violence stoppable only with a bullet. Belief in physical and mental class uniformity was an artifact of the positivism of the time.

The adoption of "man-stopping" bullets by police departments worldwide has a tortured history only a little more discernible than the prison adoption. This is more evident in the debate over the means used by police to subdue suspects and quell public disturbances, which might be protests, mass actions or strikes. Arguments in this debate often turn around a proposal by police officials to adopt expanding bullets, rarely described as "dumdums" but often as soft-point, hollow-point or not fully jacketed bullets. Critics call them dumdums.

The firearms and bullets used by police are determined by policy and not by individual choice, and policy is decided by elected officials. Individual officers may feel pressure to carry bullets deemed more likely to stop persons of lower nervous organization (or however it is put) than the bullets policy allows.

When the dumdum technique first became widely known and its utility against the raging other publicized in the news of battles won, a soldier or a police officer might apply it to standard issue low caliber bullets to give himself confidence under trying conditions. A prime example is the Corporal of the Marines who published his diary of events during the siege of the legation quarters in Beijing, June-August, 1900. Chinese nationalists called "Boxers" by English speakers as well as elements of the Chinese Imperial Army forced foreigners to seek refuge in the quarter until the siege was raised by troops of an eight nation coalition. The Corporal was an English South African who arrived with a contingent from the colonies.

---

[208] Prison Riot Stopped, *Dakota County Herald,* June 14, 1917: 6
[209] Esposito and Wood (1982: 172)

On June 15 he wrote in his diary that it would be good to have dumdums and he intended to make some. On July 3 he shot a Chinese "brute" with a "homemade dumdum."[210] He does not refer to dumdums again, but he did have other occasions to shoot at the enemy. On his return to South Africa he became a Constable in the Urban District Police of Cape Colony. He no doubt retained the skills to remake bullets and experienced situations under which it would seem in order to do so.

The transition from acquired individual ability to make dumdum bullets to the need to arm protectors of the peace with less bluntly named projectiles is a matter of greater media attention.

The murderously persistent attacker underlined the need for police to be armed with projectiles that halted an assailant with one shot. The death of Constable Joseph Macdonald of the Sudbury, Ontario (Canada) police force on October 7, 1993 was offered as an example of what could be avoided if police revolvers contained hollow-point bullets rather than the straight jacketed bullets two of which passed through a man MacDonald was attempting to subdue before the man returned fire.[211] MacDonald's death became a cause celebre, lending his name to a memorial fund, a bridge and a sports team. The circumstances of the young officer's death induced an effect similar to that exercised by the reports of the onslaught of the wounded ghazi or dervish one hundred years earlier.

There were few instances as resonant as MacDonald's. The case for expanding bullets in police arms was more often made by reference to the image of the attacker, which was extended yet further to encompass "the pyschotic, enraged or chemically intoxicated attacker who is oblivious to being shot."[212] The criminal anthropology that allowed anyone to become a candidate for a bullet of enhanced stopping force had become the predominant likelihood given the lack of self-control of the criminally insane and the widespread consumption of drugs.

The concept of a margin of safety enters into this advocacy of a bullet that expands on entering a body and extends sharp metal claws from its peeled back copper nose jacket. The spinning projectile has a slightly greater likelihood of ripping a vital blood vessel or organ than a plain expanding bullet. That potential might be sufficient to halt a crazed drug fiend who would press forward after being shot with a police .38.

This type of bullet, called the Black Talon and other names, was promoted as giving a greater possibility of a one-shot stop, a likelihood sufficiently greater to adopt the bullet in the face of the negative publicity it was certain to cause once its operation was made known. If the image of the unstoppable attacker was used to justify expanding bullets, the image of the ever more unstoppable attacker called for ever more deadly bullets ever more carefully disguised.

The manufacturer's description of the Speer Gold Dot bullet, for instance, emphasizes both its expanding and penetrating power. If a bullet expands too rapidly or fragments on entering the skin it may not reach the deeper blood vessels or vital organs. A photograph of the expanded bullet shows its lead core branching out in four uniform blade-like extensions which would form as the bullet plows ahead. The trademark gold dot is

---

[210] Corporal of the Marines (1900: 211; 213)

[211] Kaye (1996: 197)

[212] Firearms Tactical Institute (1998)

at the center of the expanded lead core. The manufacturer's literature concentrates on the special process bonding the core with the jacket to prevent separation of the two, and the consequent failure to expand.

Swedish police replaced their Norma Sak fully jacketed bullet with the Gold Dot as the service bullet in 2003. Civil rights activists filed a lawsuit against the national police in 2012, charging that the shooting of a mentally ill man in his home the previous year would not have been fatal if the Gold Dot was not the bullet used.[213] As in many other cases it was not clear whether police needed to stop the man in the way the Gold Dot could, or if another bullet would have done as well without killing the man.

Manufacturers did not use the phrase "dumdum" in bullet specifications. It only was applied from the outside as a term of condemnation. A rhetoric of concealment that also assured police officials that the bullet had stopping power became the language of bullet presentation those affected attempted to rewrite in the language of the dumdum complex.

The perforating power of the existing small caliber cartridges fit into official language as a reason to switch to the penetrating but not perforating expanding bullets. The danger pass-through bullets posed to bystanders became both a technical and an anecdotal argument for abandoning the smooth, unchanging projectiles in favor of bullets that grew and stopped in the body they were aimed at.

A coroner's inquest ruled that 39-year old David Sycamore was "lawfully killed" by police fire on the steps of Guildford Cathedral when he failed to comply with police orders to drop a gun he was aiming at them.[214] Sycamore's gun turned out to be a blank pistol, and he was a chronic depressive who had said that he would make the police kill him. These factors set into relief the tragic circumstances of the man's life. Examination of the scene afterward revealed that both of the bullets fired at Sycamore passed through his body and lodged in the cathedral structure behind him. They were hard-shell bullets uninhibited by any deformation. They might just as easily entered someone in the cathedral. That possibility was used to promote police switchover to expanding bullets.[215] The vulnerability of religious faithful at church to stray police bullets added to the protective appeal of high stopping power.

As with all gun control initiatives, high profile cases of individuals dramatized the police need for expanding bullets and the dangers expanding bullets posed to innocent victims. The exchange of charges associated with military use of the bullets had ceased with the advent of asymmetrical warfare and remotely piloted systems. The dumdum complex centered around police ammunition type. The police had also inherited the lower nervous organization targets of the colonial military, but those targets were not so easily portrayed as outsiders posing a threat to the community as a whole. They might be mentally ill people brandishing replica guns posing no genuine threat to police or bystanders.

This changed the nature of those potentially stopped by expanding bullets. Anyone might become a criminal. The wellsprings of public threats were more varied and less culpable than the criminality of moral fault with physical manifestations. The array of reasons that Ellis listed for criminal behavior were less grounded in absolutes of appearance and body constitution than the old descriptions of savages. Mental

---

[213] Hårdh (2012)

[214] Man shot by police outside Guildford Cathedral "lawfully killed," *The Telegraph*, August 20, 2009 [online]

[215] Higginson (2009)

illness and drug use, stress and deprivation are among the factors that figure into what formerly was classified as criminal behavior. Police authorities ask for bullets with more stopping power and contend they are protecting the crowd from the projectiles that might pass through the target who will continue on his course to kill the shooter. Opponents make a case for the inhumanity and illegality under international law of what they do not hesitate to name dumdum bullets.

It is in the interest of advocates and critics to emphasize individual cases both of police officers who might have been saved if the assailant had been forced to stop and of victims who should not have been killed to subdue them. Since these disagreements began with the invention of dumdum bullets, there has been a two-sided study of wound ballistics part experimental and part observational. Physicians have long observed the pathology of bullets in flesh. Ballistics experts, dependent upon physicians for the anatomy, have devised a physics of the bullet in the body in order to evaluate the worth of one bullet or another. There is no level of observation that tracks the precise effects of any one bullet type in every possible body and under every possible set of conditions.

Wound ballistics as distinct from other less intimate ballistics, exit ballistics, trajectory ballistics, and terminal ballistics, is as subject to theorizing as any branch of physics, but unlike the other areas of ballistics it is less susceptible to experiment. Living and dead animals as well as human cadavers have been used for wound ballistics studies. Since the debate over expanding bullets centers around stopping power these methods have an obvious deficiency. How does a bullet stop a person intent upon harming the shooter? How can stopping power be measured to assess the relative stopping power of different bullet types? These questions can only be approached theoretically, generating hypotheses tested in abstract experiments. And by recursion to anecdote.

The second of a decisive confrontation, so well exploited by visual media, remains a picture to be differentiated ballistically. Various systems of measurement have been proposed; composite indices that include physical characteristics of the shot and of the recipient, have been assembled. One proposal has been the "one-shot stop" estimates garnered by tabulating type of bullet used by the police against the persistent attacker.

A former patrol officer with the Detroit, Michigan police department, Evan Marshall, set about collecting instances of suspects and assailants shot by police, shop owners and others in defensive shootings. Collaborating with Benton County (Indiana) sheriff's deputy Edwin Sanow, over a period of fifteen years Marshall accumulated data on over 1800 instances of first hits to the head or body.[216] In each instance he noted the caliber and type of the bullet, bullet weight, velocity and whether or not the shot stopped the target. A one-shot stop was defined as "instantaneous incapacitation" or "incapacitation after fleeing not more than 3 meters." Marshall and Sanow calculated the stopping power of a bullet as the ratio of total shootings with that bullet to the number of number of one-shot stops. Theirs is the viewpoint of the patrol officer as barricade defender assessing the effectiveness of the shot and not the reasons for making it.

Marshall and Sanow introduced their findings in contrast to other numerical analyses of bullet stopping power: the Relative Incapacitation Index, which relied on the results of soldiers firing Colt .45 handguns into

---

[216] Marshall and Sanow (1992)

fixed targets and into ballistic gelatin, and the Strasbourg studies, which measured the extent of wounding in live goats. Marshall and Sanow's tables were based on "street results," on the study of actual shootings represented in the reports and recorded observations of police officers, physicians and other medical personnel who were interviewed by the researchers. The findings do not rely on anecdotes, but are permeated by them as illustrations of one or another parameter. From the outset the authors deny that there is any such thing as a certain one-shot stop, a caliber or bullet type that guarantees the target always will stop. It rather is a matter of likelihood expressed as a percentage ratio for each bullet, which requires discovering what increases that ratio.

The .38 Special 158-grain bullet, a round-nose lead type bullet commonly issued to police in the United States between the 1920's and the 1990's, when fired with a velocity of 704 feet per second achieved 160 one-shot stops in 306 shootings, for a percentage of 52.28. The same caliber but a lead hollow-point type expanding bullet, same weight but fired with a higher velocity, 926 feet per second, improved the percentage, 79 out of 114 total, or 69.29 percent. A jacketed hollow-point (JHP) fired at a slightly greater velocity did not increase the percentage, nor did a lighter 0.38 Special jacketed hollow point fired at a much higher velocity.

Caliber, bullet type and velocity seemed to have the greatest effect on the percentage of one-shot stops among the parameters surveyed. The 9mm parabellum 115-grain jacketed hollow point bullet at 1304 feet per second greatly exceeded other 9mm PB JHP bullets at lower velocities in the percentage of one-shot stops (89.47 percent). The .357 Magnum 125-grain jacketed hollow-point at 1453 feet per second had a higher one-shot percentage (96.96) than a .357 Magnum 158-grain JHP at 1233 feet per second and a .357 Magnum 125-grain JHP at 1391 feet per second. A medium weight .357 Magnum at high speed had the greatest stopping power.

Marshall and Sanow concluded that street results favor the medium weight .357 Magnum JHP for police use. Their evaluations drawn from actual shootings had some influence on the trend toward the adoption of expanding bullets by police departments, though the departments themselves often based their public demand for change on an emotion-charged shooting reported in the media.

The one-shot stop methodology and results seem geared to overcome the objections raised against the published "street" studies of bullet performance. Recognizing that this manner of study has an "allure" that transcends laboratory and clinical research, an FBI special agent in the firearms tactical unit reported in an 1989 paper that the existing research on shooting incidents "is collected haphazardly, lacking scientific method and controls, with no noticeable attempt to verify the less than reliable accounts of the participants with actual investigative or forensic reports."[217] Marshall and Sanow tried to answer these objections with a large sample size, perusal of reports and interviews with participants and researchers.

Some of the criticisms that followed Marshall and Sanow's 1992 publication nonetheless found deficiencies in these areas, and a few questioned their veracity in reporting the data. But the most serious deficiencies in their work also were anticipated by the 1989 paper: "cases are subjectively selected (how many are not included because they do not fit the assumptions made?). The numbers of cases cited are statistically meaningless, and the underlying assumptions upon which the collection of information are based

---

[217] Urey (1989: 13)

are themselves based on myths such as knock-down power, energy transfer, hydrostatic shock, or the temporary cavity methodology of flawed work such as."

Someone falling may rise to fire again whatever the belief about what the bullet has accomplished internally. Those cases would not have been included in the survey. By the same token, the targets actually felled may have been targeted by expert marksmen who understood that the brain or a blood vessel has to be hit for there to be true incapacitation.

Those who accept the myth that a one-shot drop ends the career of attacker may not survive that myth.[218] The tension between the force of the bullet and the force of the onslaught, first recognized in the invention of Puckle's Defence, is still covered by a mythology, that the effectiveness of the defense can be determined in the design of the bullet alone. No matter how well planned, the bullet still did not take shape. All of the parties recognize the ever growing category of the attacker. "Those individuals who are stimulated by fear, adrenaline, drugs, alcohol, and/or sheer will and survival determination may not be incapacitated even if mortally wounded." The lower nervous organization may be induced or natural.

The attempt to construct an inescapably effective bullet grows more urgent with the broadening of the categories of attackers, who seem to be the enemies of order and of safety itself. The answer of order and safety to this threat is to seek a bullet that always will fell its mark. The folkloric magic bullet that always strikes its mark is not sufficient; the bullet has to incapacitate its mark as well. Belief in that incapacitation with one shot is itself a folktale but one with pretensions to science. The latest seemingly technical proposition for even a predictably effective bullet is called into question by hard facts and by the manner in which the hard facts that support the effective bullet have been collected and analyzed.

This image of a community defended from destruction by a projectile relaxes the stress of being beset by threatening forces, whether animals or humans of lower nervous organization. It makes reaching the other side of the moment of extreme fear a foregone conclusion. The attacker will be knocked down, blown away, taken out, dropped. The telepathic force of the community will be focussed in a sure-fire missile. The one-shot stop and its critique together form a discussion of threat response, a shared endocrinology of the violent defensive surge.

The critique of Marshall and Sanow does not deny that the threat exists; it dismisses the one-shot possibility as dangerous overconfidence. One-shot oversimplifies the enemy; critique elaborates the enemy. The multiplicity of bullet types, calibers, weights, velocities and other measures of shape and energy, the result of technological and historical evolution, many of them designed to achieve greater stopping power, among them provide a ground for endless consideration of bullet effectiveness, humanity, expense and purpose.

Small wonder that there was an attempt from early in bullet history to gear bullet design to the evolutionary level, animal nature or nervous organization of those being shot. The one-shot stop debate sidesteps characterological generalizations about the target except to envision the invulnerable attacker. Within that generality characterizations looking back to dumdums can germinate. The raging assailant might also be a member of a troublesome ethnic group or an anti-government protester. The flexibility of the

---

[218] Pinnozotto, Kern and Davis (2004)

boundaries of "criminal" or "insane" already is apparent. The statement by one student of wound ballistics reaches into this territory.[219]

> It is well-nigh impossible to measure the effectiveness of rifle or handgun
> ammunition objectively and in a manner that is generally valid.

The lack of objectivity derives from a prior commitment to determining the nature of the party being shot.

The savage may be the target of the dumdum bullet or its shooter. From the viewpoint of expanding bullet advocates, of course, there are many more names for the savage and for the bullet. For opponents it is much simpler: whatever the bullets are called they act like dumdum bullets and whatever the targets are called, if anything at all, they're considered savages. The reminder that dumdum bullets are illegal under international law usually brings the response that the bullets being used are not dumdum bullets or that this is an internal conflict to which international law does not apply.

The specific critique of stopping power measurements can rise to a critique of expanding bullets in general, but there it encounters the questions that stopping power measurements seem to resolve.

So do we use no bullets at all?

Ignoring questions of the ability of the expanding bullets (or any bullets) unfailingly to fell the foe, advocates of these bullets look to one-shot stop models and theories. After their 1992 publication, Marshall and Sanow issued from the same press (located in the Gunbarrel district of Boulder, Colorado), two more books updating their evaluations and offering reviews of newly issued ammunition.[220] Both emphasized the stopping power of bullets.

A physician's review of the second book reiterated and extended the objections to the statistical validity and veracity of the first, restating plainly that the authors were fixing their results to favor fast, light bullets manufactured by companies to which they have connections.[221] Most police departments had not abandoned slow, heavy bullets on Marshall and Sanow's recommendation. Evidence that the one-shot stop measurements are biassed to favor bullet makers' business supplies an ulterior motive for elevating the findings for one type of bullet over others.

The bullets police departments chose to arm officers, when department officials made that choice as a matter of policy, went through an evolution from the introduction of dumdum bullet technology and nomenclature onward. There always was a pull toward issuing the defenders of urban order bullets to stop the savage onslaught whoever the savages may have been. Within that gravitational field the debate over the effectiveness of the bullets for that purpose played itself out, always anecdotal, tending toward the scientific and the pseudoscientific by turns.

During the tense period after Pancho Villa's March, 1916 cross-border raid on Columbus, New Mexico, the Los Angeles Police Department prepared for upheaval among Mexican-Americans who supported Villa's revolutionary aims. One of the expedients adopted was to issue dumdum bullets to the police.[222] It is

---

[219] Kneubuhl (2011: 176)
[220] Marshall and Sanow (1996) and (2001)
[221] Fackler (1997)
[222] Escobar (1999: 74)

not clear whether or not the police ever shot Mexicans or anyone else with these bullets. The gesture clearly was motivated by an American population governing land that once had been Mexican treating the dumdum explicitly as an instrument of halting a possible uprising.

The anthropologist Daniel Nugent during his research into the revolutionary present and past of the community of Namiquipa, Chihuahua (Northern Mexico) learned that the son-in-law of an elderly woman home owner dug a dumdum bullet from the adobe wall of the house after federal police had gassed and fired upon participants at an opposition party demonstration nearby.[223] The official police policy, Nugent added, was to load only blank cartridges when controlling political rallies. This incident took place in the early 1980's which Namiquipa residents viewed as another time of revolution following the 1910-20 period. In both 1916 Los Angeles and 1980's Namiquipa the police resorted to dumdums against the outsiders, Mexicans in Los Angeles and opposition party members in Namiquipa. In Los Angeles the gesture of dumdum distribution was made known and served as a confidence builder to the police. In Namiquipa it was concealed and may even have been a spontaneous alteration by the police officer who fired the bullet. In neither case was there any claim of testing the dumdum bullets. Their reputation was sufficient to preserve the division between would be shooters and their victims.

The early 1990's were a time when police departments considered bullet choices based on criteria like those put forward by Marshall and Sanow, allowing them to package safety and effectiveness considerations together while concealing the savage target of the projectile. The Los Angeles Police Department explicitly adopted dumdum bullets in the face of the threat of gangs of varying ethnic composition though it was as much an act of defiance against dumdum critics as against the gangs.[224]

In response to an accusatory letter in *The New York Times* in 1988, a Firearms and Tactical Section officer in the New York City Police Department denied that the department had ever used dumdum bullets.[225] In 1994 a carefully worded proposal made it clear that the City Police Department was considering a conversion to hollow-point bullets, which are not dumdum bullets, the proposal emphasized.[226] There was no allusion to the type of suspect who would require these bullets. Officer survival and public safety were advanced as sufficient justification for their adoption.

---

[223] Nugent (1993: 165n1)
[224] Hainey (1991)
[225] Cerar (1988)
[226] Krauss (1994)

11. Gangster Dumdums

The softest bomb the Luftwaffe dropped on Britain during late 1940 was a leaflet picturing then recently elected Prime Minister Winston Churchill, bowler on his head, cigar clamped in his mouth and a Thompson submachine gun held at ease but ready in his hands. The photo had been taken during Churchill's morale boosting July 30 visit to coastal defense installations. Two soldiers standing near him had been removed for the newspaper published photo. On August 11 Churchill took part in ordnance practice near Chequer, where he was quoted saying, "the best way to kill a Hun was to use a snub-nosed bullet."[227] "Snub-nosed" referred to dumdums with cut tips. Churchill's son Randolph protested that the bullets were illegal in war. The Germans by their savagery had forfeited any consideration due the civilized in warfare, was his father's reply.

38. German propaganda leaflet, 1940.

---

[227] Manchester and Reid (2012: 189-90)

The text of the leaflet, composed as a wanted poster, compared Churchill to a gangster who murdered women and children. The Thompson sub-machine gun, the "tommy gun" Churchill was holding was a prop familiar to audiences from American gangster movies, which pictured large, noisy, rapid fire guns mowing down groups of people. Even today my first sight of Churchill holding the gun set up a discordant equation with an antiquated mob gunman, so well has the gangster-gun meme been implanted in the image bank.

Joseph Goebbels intended this leaflet to identify the prime minister as a lawless murderer. Instead, Churchill came off looking like a defender of the realm against the barbarian hordes. He cultivated that image with public statements that he was ready to fight the invaders to the last. He was no gangster but a freeborn Englishman with native firepower. The intensity of national defense called for ammunition of a special efficacy against savage attackers.

Whether or not gangsters fired dumdum bullets (or for that matter, tommy guns), the public at large so strongly associated the bullets with them that Churchill could evoke it for his own propaganda purposes by holding a gun and making a remark. The Nazi propaganda machine supposed a popular semantics for the machine gun pose that could turn the image and remark against him. National dumdum charges and countercharges were much publicized by newspaper and magazine articles from the wars of the early 20th century through World War I, when they reached a peak. Gangster dumdums did not receive press until the appearance of celebrity gangsters in the 1920's. Churchill's assertion and its attempted remaking arose from crossed readings of guns and dumdums. Churchill looked back to the colonial defense in which he matured; the Nazi propagandists looked to the gangster frame.

In 1899 Churchill was in the Transvaal, South Africa reporting on a British military action against the Boers when he was captured. General Botha saw him trying to ditch a clip of Mauser ammunition he was carrying: he explained that he had just picked it up.[228] If Botha had inquired further he would have discovered that it was Mark IV and V bullets Churchill brought with him from the battle of Omdurman in the Sudan. They were the manufactured version of dumdum bullets of the type Churchill knew full well had been rendered illegal by international agreement at the Hague conference earlier that year. Churchill was released and returned to writing his column for *The Morning Post*. The following March he denounced the Boers for using "expansive bullets" in the conflict. "A man, I use the word in the fullest sense, does not wish to lacerate his foe, however ardently he may wish his death." Yet Churchill also found the Boers to be considerate adversaries in their treatment of captive soldiers and only supplied them with expansive bullets for the bad reputation associated with them.

Earlier still, during the era of dumdum discovery, Churchill had taken a different view of the bullets. A young cavalry officer in India, he took part in the 1897 Malakand expedition by the British army against Pashtuns on the northeastern frontier. He already had begun his career as a journalist and author of first-hand campaign histories. In his writing on the Malakand campaign he took up the topic of the bullets unofficially called "ek-dum" ("at once" in Hindi).[229] He referred to them as "expansive bullets" and wrote that

---

[228] Manchester (1983)
[229] Churchill (1901: 287-89)

their "stopping power was all that could be desired." He maintained with other army officers that the "Dum-Dums" were not in violation of the St. Petersburg and Geneva conventions.

> I would observe that bullets are primarily intended to kill, and that these bullets do
> their duty most effectually, without causing more pain to those struck by them, than
> the ordinary lead variety. As the enemy obtained some Lee-Metford rifles and Dum-Dum
> ammunition during the progress of the fighting, information on this latter point is
> forthcoming. The sensation is described as similar to that produced by any bullet-a
> violent numbing blow, followed by a sense of injury and weakness but little actual pain
> at the time. Indeed, now-a-days, very few people are so unfortunate as to suffer much pain
> from wounds, except during the period of recovery.

In language that approached the "humane bullet" wording of some of his contemporaries, Churchill minimized the pain felt by the recipient of the bullet after reflecting that more would be learned on the subject now that the enemy has obtained rifles and dumdums. He shared this nonchalance with fellow journalist Ernest Nathaniel Bennett, who had decried the atrocities committed against the dervishes at and after Omdurman[230]. In his account of the campaign Bennett justified the use of expansive bullets as the only way to stop "the far less delicate organism" of the onrushing savage.[231]

Churchill didn't justify the use of dumdums against the savages with a low/high nervous organization dichotomy as Bennett did. He distanced himself from the effects of the bullets without erecting that buffer. Holding the tommy gun in the doctored newspaper photograph and subsequent propaganda leaflet he had situated himself on the barricades of an England about to be invaded, needing only to place the Huns in the savage category to warrant the savage ammunition. He had moved one step further back from condemning the Boers to the bullets he said they used. Churchill was present in the environment of the dumdum's invention and was conversant with their political uses as accusation and as threat. Forty years later the imagery of gangster dumdums intervened.

The earliest association of dumdums with gangsters in the United States, where they were most noticed, emerged in the statement of Antonio Comito, a member of the Novello family of New York on trial for counterfeiting. In February, 1909 Ignazio Lupo instructed confederates in the technique of cross-slitting the noses of the bullets he brought to them as part of a gun and ammunition delivery.[232] There was no follow up information about where Lupo learned this skill, or how much more effective than the unmodified rounds these bullets seemed to be. It was at least a confidence-building gesture as it had been for hunters facing dangerous game or troops facing unstoppable enemies in the colonial wars. The newspapers and journals did not report the trial or the armaments of the accused counterfeiters. Dumdums were restricted to news of military exchanges.

The arrival of the Mafia and organized crime as categories of journalism in the postwar period created a point of attachment for the dumdum complex. The bullets were fitted into any description of a death

---

[230] Bennett (1899a). Churchill initially supported Bennett's views but then expunged critical material from the second edition of his book on the Sudan campaign not to appear too critical of national wars.
[231] Bennett (1899b: 229)
[232] Dash (2009: 176). Antonio Comito "statement in re Morello-Lupo case" August, 1910, Lawrence Richey Papers, Herbert Hoover Presidential Library

resulting from gang action without reference to an autopsy or a scene investigation. On July 23, 1921 "Two Gun" Johnny Guardino was caught off guard and his body was "riddled with dumdum bullets," the eleventh political murder in Chicago's "bad lands."[233] Guardino was a hit-man charged with avenging the death of Tony D'Andrea, union leader and president of the Unione Siciliana, in the midst of a war for control of Chicago's nineteenth ward. The Irish boss of the ward, Johnny Powers, was struggling to retain his position as alderman where the Sicilian population had exceeded the one time Irish majority. The dumdums typically were aimed across tribal lines only this time they passed between two urban ethnic groups in conflict over territory. There was no "us and them" with the "us" claiming to be shocked at the use of dumdums. There was only the audience.

On September 7, 1928 another president of the Unione Siciliana, Antonio Lombardo, was gunned down with dumdum bullets, according to news reports.[234] The nationwide reach of the gangs and their fights was made plain by the suggestion that the killing of Lombardo and his associate Tony "The Pelican" Ferrara was retaliation for the murder of Frank Uale (Yale) in Brooklyn, New York the previous July. Another of Lombardo's guards present at the time was not shot with dumdums like Lombardo and Ferrara, probably because he was not implicated in the Yale shooting.[235]

Al Capone, an organizer of crime, generally did not appear holding a machine gun but his name cropped up in accounts of the Lombardo shooting as an ally of the dead man and the likely author of the Yale shooting. A police raid on the apartment of Capone's henchman "Machine Gun" Jack McGurn in 1928 unsurprisingly yielded a Thompson submachine gun "loaded with a clip of 50 dumdums notched at the tip."[236] The technique that Lupo imparted in 1909 had become a gangster tradition. Deaths from dumdums and dumdums discovered more or less followed the same pattern as wartime accusations with the gangsters the stand-in for the universal enemy. "Two high powered rifles loaded with dumdums" were among the cache of weapons found in the lodgings of two apprehended Florida gangsters.[237]

Dumdums remained in the stream of gangster reportage, adding an atmospheric detail in fiction and news but no more. New York city policeman Leonardo Grossman was heard on a tape recording telling Michael Scandifia that he would obtain dumdum bullets for Scandifia's revolver that would "blow a hole this big" in stool pigeons.[238] The gangster facility with dumdum conversion had passed to the police, who were officially experimenting with them at the time. Finally, the bullets were available to all for purchase.

Count 32 of an indictment handed down by a New Jersey grand jury on May 14, 2010 against a member of the Lucchese crime family read

SAMUEL A. JULIANO on or about December 6, 2007, at the Borough of Glen Ridge, in the County of Essex, elsewhere, and within the jurisdiction of this Court, knowingly did possess dum dum bullets, that is, one box of fifty, .357 Magnum 110 GR jacketed hollow point bullets, contrary to the provisions of N.J.S.A. 2C:39-3f, and against the peace of this State, the government and dignity of same.

---

[233] Woman Twelfth to Die in Chicago 'Bad Lands,' *The New York Tribune,* July 25, 1921: 1.
[234] Pasley (1930: 231)
[235] Lombardo Aide Dies from his Wounds,*The New York Times* September 10, 1928.
[236] Gusfield (1928: 146)
[237] Parish (2008: 94)
[238] Department Trial of Grossman Ends, *The New York Times* October 17, 1968

The statute cited in the indictment reads as follows

Except as authorized by statute, it shall be unlawful for any person knowingly to possess any (hollow nose or dum-dum bullet) [or] (body armor breaching or penetrating ammunition).

and is followed by paragraphs of annotation stating exactly what a hollow nose or dum dum bullet is and what it means to possess one.

Exceptions written into the statute allow "sportsmen" to possess, store on their property and transport hollow nose bullets from the point of purchase and to a firing range. That the statute is intended to distinguish between classes of firearms possessors was made clear by the 1976 decision of the New Jersey Supreme Court: the law is designed to prevent "criminal and other unfit elements from acquiring firearms while enabling the fit elements to obtain them with minimal burdens."[239]

Juliano and the gangsters of past and present were unfit elements of society according to the schema of the statute as interpreted by the court. Unlike the fit elements, sportsmen and police going about their duties, they were not to be permitted to own or transport dumdum bullets. Churchill, in his role as defender of the nation, would not have been indicted for possessing dumdum bullets in 2010 New Jersey.

By defining dumdums according to statute, states like New Jersey have perpetuated the post- World War I assignment of dumdums to gangsters. The broader category of "criminal and other unfit elements" sets out boundaries that might include the mentally ill among those who cannot legally possess the bullets. The statute criminalizes someone otherwise judged to be unfit found to be in possession of the bullets. They could only be prevented from purchasing them by a system of background checks that discover a history of behavior rendering the purchaser unfit. The extent and applicability of such a system, and the ability to enforce it, are currently undetermined.

Dumdums as concrete objects have precipitated virtually unnoticed into present debate over firearms in America. But any jacketed bullet can be turned into a dumdum, as the news and cartoons of earlier in the century illustrated.

---

[239] Service Armament Co. v. Hyland, 70 N.J. 550, 559

12. Assassinations

On the night of May 3, 1906 a party containing Herbert Munro Stainbank, the magistrate at Mahlhatini, was fired upon as Stainbank made his way to a telephone station to report to his superior on the unsettled conditions in that area of Zululand. One of the police escorts was struck and Stainbank himself was hit in the knee, causing a wound from which he later died. During July of the following year five Zulu men were placed on trial for planning and executing the attack.[240] The Crown prosecutor contended that Umpeta, one of the men charged, was hired by the other four men to shoot the magistrate. The prosecutor asserted that Stainbank was targeted because he supported Sengungu, the acting chief of the district, and refused to relinquish the position to Makowkanka, who claimed to be the rightful successor.

Umpeta was a follower of Dinuzulu, the Zulu king, as the *tshoteobezi* (glittering ornaments) hanging from his bandolier announced. The other four defendants were of Makowkanka's party, but Umpeta was not. He was hired for his skilled marksmanship.

One of the prosecution's articles of evidence was that Stainbank had been shot with a dumdum bullet, apparently found in the wound though there was no comment on that in the trial transcript. Umpeta cut the ends of his bullets to flatten the noses and expose the core "so that when he was out hunting his shot on striking a buck would expand." A wound in the leg joint was the hunter's way of immobilizing an animal that otherwise could run some distance with a bullet in its body.

Umpeta was reported to have hidden his bullets when he returned to his dwelling after the shooting. Asked to send his gun for inspection, he only sent an old blunderbuss, not the Lee-Metford rifle he was known to possess and use. That and other evidence was not sufficient to convict him of the crime of murder. The jury found Umpeta and the other four men not guilty.

The trial of Stainbank's alleged killers took place shortly after the last significant rebellion of a Zulu leader against British authority ended in the battle of Mome Gorge on June 10, 1906. Leading this protest against taxes and land seizures, Bambatha, the chief of the Zondi, a Zulu offshoot, was killed in the battle. Dinuzulu, who had returned from 7 year exile on St. Helena in 1897, vigorously denied fomenting the rebellion but was sentenced to 4 years imprisonment. The Active Militia formed by white colonists in Natal used Maxim guns firing dumdum bullets to cause heavy casualties among the assegai and shield-carrying warriors.[241]

Charles Maxwell, a missionary who visited Zulu settlements in 1906, illustrated the "witchcraft delusion" of the Zulu in his account of a confrontation between Zulu followers of a priest who had granted them immunity to bullets, and a formation of Natal soldiers.[242]

> A few days later each man carrying only one assegai, because that one was all
> it was supposed he could need, walked into the faces of an equal number of Natal
> soldiers whose rifles repeated ten dumdum bullets per minute for each soldier,
> and a number of newly invented Maxim guns, able to discharge 600 shots per minute.

The outcome was disproportionately in favor of the Natal soldiers.

---

[240] Great Britain. Parliament. House of Commons (1908: 221-32) contains the report of the trial proceedings.
[241] Marks (1970: 162)
[242] Maxwell (1906: 590)

In the midst of an environment of oppression and revolt in which the revolts are put down by soldiers armed with repeating rifles shooting dumdum bullets a Zulu man shoots at a colonial administrator's party. He fires the flat-nosed version of dumdum bullets he has made himself by cutting off the noses of conventional .303 rounds, Churchill's "snub-nosed bullets." This is a practice of big game hunters adapted to assassination. The object of the attack dies because a bullet lodged in his knee causes a fatal infection. Zulu men with appreciable motives are charged, but no one is ever convicted by the colonial judicial system.

This might seem to be the revenge of the oppressed against an individual of the occupying power. It is as much the application of hunting skills to a political action. British missionaries may have deplored the delusion that led Zulu men to their deaths carrying spears against dumdum armed soldiers. The assassination demonstrated that the colonists were not protected from the force of their own bullets fired by their own rifles if it came to that.

Originating in the use of deforming bullets in hunting and as a confidence builder for soldiers sent out against the savage natives, the bullets are adapted to political purposes when used to target important individuals in assassinations. Whether they are any more effective than other bullets or even other weapons is never decided by these instances.

The Zulu had become enmeshed with the Boers when they were hired as mercenaries in the 1860's to war against the British colonists. The Boers' demands for compensation in the form of land concessions ultimately were settled with the provision of farm sites in an area that later was absorbed by the Transvaal state. The battle between the Boers and the British Empire was waged alongside Zulu revolts against the British until the culminating warfare and treaties of the early twentieth century. The Zulu were at first the savages addressed by the dumdums, then were incorporated into the mutual dumdum accusations of the armed nations such as the Boers and the British.

The Zulu mostly abandoned military action and took part in the formation of a series of organizations culminating in the African National Congress in 1911. Natal was a province of the Union of South Africa from the nation's formation in 1910 until 1994 when it became kwaZulu Natal province.
Dumdums were a bilateral political tool anticipating another ultimately more effective one.

Assassinations where dumdums are said to be used are a sign of emergent engagement of dominant with rebellious adversaries in an accusation structure of warfare processed through the courts and on an informal battlefield. Dumdum accusations give a bullet shape to the politics of assassination.

Ito Hirobumi, referred to in Western media as Prince Ito, was assassinated at the Harbin railroad station in Manchuria on October 26, 1909. Ito, the governing military official in Japan-occupied Korea, had come for a meeting with the Russian finance minister when a Korean nationalist fired six shots, striking him three times. Three other Japanese officials standing on the platform nearby also were wounded but the Russian minister was not. Ito died not long afterward, in his last words calling his assassin a "fool."[243]

The assassin, whose name is given in different Romanizations in European language newspapers, was apprehended on the site, shouting in Russian, "Korea oora," "Korea, forever."

---

[243] In an admittedly third-hand version of the story Henri Mylès recounts that Ito's last words referred to Koreans generally as fools, who did not know what they had done to themselves by killing him. Mylès (1913: 47)

An Jung-geun was relieved of his weapon, an FM Browning N1900 semi-automatic pistol chambering .32 ACP rounds. He was escorted into detention and after a trial, during which he demanded to be treated as an enemy combatant and not as an assassin, he was executed by hanging, a criminal's execution. Chinese, Korean and even some Japanese public figures openly sympathized with him, possibly because he espoused a pan-Asian philosophy and listed as one of his grievances against Ito that the official had ceased to follow the Emperor Meiji. Despite the ban placed on ecclesiastical last rites by the Catholic Bishop, An was given solace by the priest who had converted him to Catholicism in his youth.

The newspapers gave varying but not contradictory accounts of the assassination. Some included more details than others. Japanese, Chinese, European and American newspapers had correspondents present in Harbin to cover an important diplomatic initiative by Ito, who had been the first prime minister of Japan and held the office three more times before being appointed governing general of the Korea sphere of influence.

Declaring "Nippon in the House of Sorrow," *The San Francisco Call* on its October 27, 1909 front page gave the scarcest details on the assassination itself and focused on Ito's achievements as a statesman. An had shot him in the back. The other victims were listed. The newspaper compared Ito's death with that of an American advisor to the Korean government in Japanese employ, Durham Stevens, in San Francisco on March 3, 1908, while he was walking to a ferry station. Two Korean men attacked Stevens and one shot him. The man apprehended was serving a 25-year prison sentence.

The Japanese newspapers, including *The Japan Times,* the English-language daily published in Tokyo, gave admiring but terse obituaries of Ito. He had not fit the mold of the doomed warrior that all classes of Japanese society preferred.[244] He had composed a poem: "Drunk, I (relax) with my head on a beauty's lap, awakened (refreshed), I grip the reins of power."[245] His dalliances with women were the subject of cartoons in Japanese newspapers.

The *Hawaiian Star* headed its October 10, 1909 front page with a banner not seen elsewhere: "Twenty Assassins Ready With Bullet and Poison." Immediately below another line took up the poison theme.

ITO'S SLAYER USED
DUMDUM BULLETS
MADE POISONOUS

39. Headline. Hawaiian Star, October 28, 1909:1

The text of the report used part of its three lines to qualify the poison, cyanide of potassium.

The other main English-language newspaper, *The Hawaiian Gazette,* included more detail on the assassination and also printed the poison dumdum claim.

On November 2, 1909 the *Gazette* reprinted an article from another Hawaiian English-language newspaper, the *Advertiser,* which warned that Lu Shun, the editor of a "revolutionary" Chinese-language

---

[244] In 2008 a Chinese manufacturer issued an IJA 21st Division action figure of Major Hirobumi Ito, trim and in uniform beside his horse. An inscription on the package reads: "this product is not intended to glorify or exploit the horrors and atrocities of war."

[245] Akita (n.d.)

Hawaiian newspaper, *Chee Bow Shin*, was about to be arrested together with some Korean residents of Hawaii for writing an article deploring the failure of Chinese residents of Korea to stab Ito before the hero An shot him. Lu had referred to the poisoned bullets in his piece, a translation of which was making its way to the United States district attorney's office. The monster of diplomacy, Ito, having swallowed Korea, was in Manchuria to begin the work of swallowing China.

The large Chinese and Korean communities in Hawaii were consumers of news through papers in their own languages, and had mutual interests in trying to stem Japanese aggression in the aftermath of the Russo-Japanese war. Rumors of the poisoned dumdum or the poisoned bullet aimed at Ito probably came from these sources. Recall from the previous chapter on Japanese bullets during the Russo-Japanese war that they were "poisoned" to make them germicidal, causing wounds not to become infected. Poisoned dumdums crop up occasionally in news reports. No one explains how it was known they were poisoned.

The bullets An Jung-geun fired at the Harbin railroad station were not dumdums. The .32 ACP rounds are copper jacketed light-weight, low energy cartridges carried in small pistols, with little stopping power, and best at close ranges. The FM N1900 semi-automatic was a small gun, the first produced with a slider mechanism to manually expel the cartridge, cock the hammer and chamber the next bullet. Its ease of concealment and for the time rapid action for a small weapon made it the assassin's favored gun. The Serbian nationalist Gavrilo Princip killed the archduke Franz Ferdinand and his wife with the same bullets from a similar gun (Browning N1910) in 1914. An was an expert marksman. He relied on the press of the crowd in the station to get close enough to his target so that he could place three bullets in sequence to guarantee Ito would die, then wound other Japanese officials nearby with the remaining three bullets. Dumdum bullets would not have suited his purpose.

The newspapers most attuned to the Korean cause asserted that dumdum bullets were used not because the newspapers had any clear evidence that they were but because dumdums represented a decisive act of destruction likely to kill the target painfully. The poison enhanced this intent. The dumdum story originated at some distance from the event, to give a gratifying sense of vengeance to the assassin's act. Koreans and Chinese certainly were familiar with dumdums from the suppression of the Boxers and the traded accusations during the Russo-Japanese war.

The traded accusations during World War I set the theme for dumdum charges during the conflict between British officers and Irish nationalists seeking independence. On September 20, 1920 Head Constable Peter Burke of the Royal Irish Constabulary and his younger brother Michael, a sergeant in the R.I.C., were shot in a pub in Balbriggan, near Dublin. They had stopped for refreshment, a dispute had begun and the local Irish Republican Army unit appeared. Michael survived and eventually recovered from his wounds but Peter died. The size and shape of Peter's exit wound appeared to be evidence that a dumdum bullet was used. Peter had trained members of the Auxiliaries, a paramilitary body supported by the police, who when they arrived in town from their camp nearby went on a rampage, sacking Balbriggan, burning shops and killing two men.[246]

---

[246] McMahon (2001: 92)

Much of the Irish War of Independence in the countryside was waged in the form of attacks on members of the Royal Irish Constabulary, recruited from Irish Catholics and responsible for maintaining the Crown's order. The members of the Constabulary also were a source of information on local activities, including the movements of known rebels and their supplies. An experienced constable like Michael Burke was called upon to train war veterans who served as a backup to the police as Auxiliaries. These veterans were familiar with dumdums and the charges related to them. Discovering a dumdum wound in their instructor's body was an incitement to a rampage, though Auxiliaries sacked other towns without a discovery of dumdums.

British periodicals and newspapers cultivated a background of belief that the Irish rebels made regular use of dumdum bullets in ambushing the police. One British writer alleged that the rebels "hold it to be quite as legitimate to employ them in what they term 'war' as it would be to use them for the destruction of a man-eating tiger."[247] The carryover of dumdums from big game hunting to RIC policemen was not like the carryover to ghazis and dervishes.

The division in this conflict was not as clear-cut as that between imperial defenders and attacking ghazis or between hunters and tigers, which contributed to varying reactions within the British population. On the one hand the Irish were barbarians who tore apart unsuspecting police patrols with dumdum bullets.[248] On the other hand the Irish fighters appeared to be making dumdum bullets when they actually were trimming the large caliber cartridges they had to fit their smaller caliber Howth rifles.[249]

A 1916 letter from Alfred Fannin, the director of a Dublin medical supply firm, to his brother Edward accused the insurgents of making dumdum bullets.[250] A woman who owned a stationery store in the city removed dumdum bullets she saw a police officer deposit in her daughter's doll house and followed the policemen upstairs when they continued to make their rounds, to prevent them from depositing bullets elsewhere in the house possibly leading to her arrest for aiding the rebels.[251] At each stage of the conflict a witness accused one side or the other of using the bullets or trying to pin their use on the other side.

Attempts by propagandists to associate the other side with dumdum bullets replicated the primordial give and take most recently at work in the continental war. *The Weekly News*, a sheet of invective passed off as facts, in its October 28, 1920 edition featured an article entitled The Irish Gunman. The author, intelligence officer Hugh Bertie Campbell Pollard, a firearms expert who authored several books on guns, presented a photograph of the flat nose dumdums allegedly being fired at troops and constabulary officers by the Irish rebels. This was simply to place a visual image before those who had an interest in believing the rebels were using these bullets.

---

[247] Ernle, ed. (1921: 161)
[248] Hopkinson (2002: 73-74)
[249] Gleeson (1961:202)
[250] Fannin (1995: 19)
[251] *Newsletter of the Friends of Irish Freedom*, 2, 39: March 26, 1921: 7.

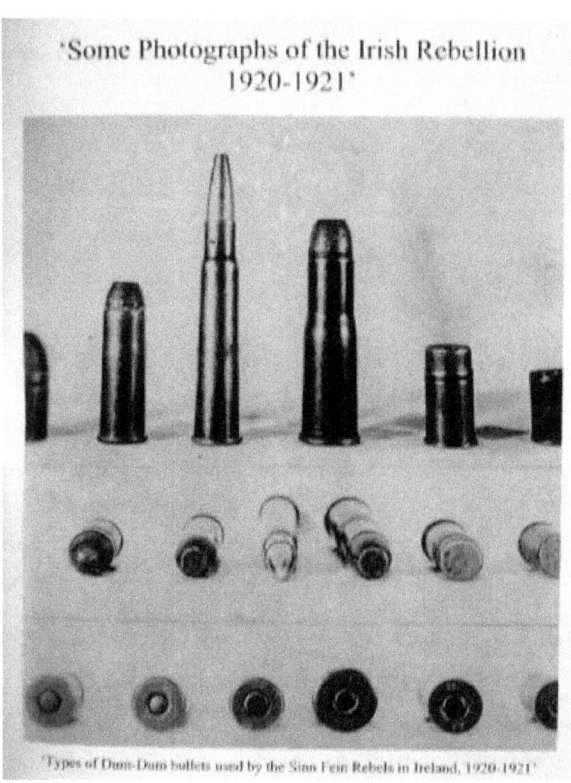

40. Types of Dum-Dum Bullets Used by the Sinn Fein rebels in Ireland, 1920-1921. Pollard (1920)

This assertion and breakdown of boundaries led to a drama of concealment and revelation, suspicions of conspiracies, sudden perceptions, violent reactions and squelched investigations. In other words, a comedy of horrors, like the one the Tory Lord Parmoor made known to the House of Lords in its April 26, 1921 session in his call for an investigation. [252]

Lord Parrow's brother, Dr. William Harrison Cripps, had written him three letters in the aftermath of the April 17 shooting Dr. Cripps and his wife survived at the hotel in Castleconnell, County Limerick, a beloved resort with an excellent view of the River Shannon. Men arrived in the hotel rather abruptly and began firing their guns and acting like "Red Indians," reducing to fragments some of the woodwork. In the distance a machine gun was heard. Though he did not see it happen it was Dr. Cripps' understanding that the proprietor of the hotel, an estimable man, had been killed. After a short time the men left in military transport vehicles, bearing with them two bodies. Lord Parrow continued,

> I have one other letter to read. I regret to have to refer to it, but I must do so,

---

[252] HL deb 26 April 1921 vol 45 cc 15-41. www.hansard.millbanksystem.com/lords/1921/apr/26/castelconnell-shooting

and I have been directed to do so by my brother. I got this letter late last night, and I went to see him this morning in order that I might make no mistake in referring to it. In a letter to me my brother says: 'I forget to mention to you I have a bullet in its cartridge case picked up by me on Sunday the 17th, the cap dented by a striker but unexploded. The bullet has been reversed, thus converting it into an expanding bullet of the most deadly character. Such bullets inflict the most terrible wounds, and were prohibited in the late war.' My brother was, of course, a great surgeon in his day. Here is the dum-dum bullet. It is not suggested that anyone fired in that hotel except the Government Auxiliaries. I do not know whether any of your Lordships would wish to see this bullet. I have shown it to two or three people who say that it is undoubtedly a dum-dum bullet, and the way it has been made such is by the familiar system of turning the point in the reverse direction.

The overwhelming conclusion of this bit of carefully presented evidence was that Crown officers in Ireland were firing dumdum bullets. In the Castleconnell shooting two separate parties fired at each other on the misconception that the other was a rebel contingent.

The clearest account of the matter had three Royal Irish Constabulary (RIC) men in plain clothes drinking at the bar when Auxiliaries, also in plainclothes, entered.[253] Each thought the others were hostiles. The Auxiliaries were part of a group, some in plainclothes, some in uniform, sent to the town to detect and capture rebels. The plainclothesmen entering the bar immediately pulled out their revolvers and ordered everyone to put up their hands, the RIC men responded with gunfire, driving the Auxiliaries out. The confused battle and deaths ensued.

No attention was paid to the absence of other dumdum bullets among those collected at the scene, nor that the one Lord Parrow exhibited was a dud. The story was echoed in newspapers and periodicals. The Friends of Irish Freedom National Bureau of Information, an American organization, encapsulated the Castelconnell shooting in their *Newsletter*.[254]

There was nothing about the RIC or the Auxiliaries that would allow one to identify the other on sight without uniforms to guide them. The usual dumdum boundaries were absent. The dumdum discovery at Castleconnell was not mentioned during the military inquiry into the shooting.

To return to the death of Michael Burke a year later, and a different type of dumdum evidence: the discovery of what they thought was a dumdum exit wound in the body must have sent a chill of ambivalence through the Auxiliaries, who also were carrying ammunition that could leave such a gaping cavity. The IRA were known to use such bullets and would kill a uniformed RIC officer. Sacking the town displaced the confusion. The dumdum markers unify people separated from each other only by uniforms, and typify the consequent violence among them.

Bullets and bodies deform in tandem where it is difficult to tell one bullet or body from another but there is a need to make a distinction. The assassin's bullet and the assassin's target combine into one changing shape.

---

[253] Dossier on the shooting including newspaper clippings on the military court of inquiry that ensued. www.theauxiliaries.com/INCIDENTS/castleconnell/castleconnell.html
[254] May 7, 1921: 5 and June 4, 1921: 4.

The death of Tupua Tamasese Lealofi III in Apia, Samoa on December 28, 1929 was declared at the Samoan National Assembly on March 5, 1930 to have been caused by dumdum bullets fired by rioting New Zealand police.[255] Other accounts placed Tamasese among Samoans struck down by machine gun fire from a nearby police station without stating the bullets used.[256] Though in either case Tamasese was killed, the accusation of dumdum use did not appear in official accounts of the events.

At the time of the shooting Tamasese was leading a Mau protest demonstration. He and eleven other protesters were victims of the attack. Eight died on the spot and three more of their wounds afterward. Estimates place the number of injured at 50. They fell in front of the courthouse in Apia, the capitol of Western Samoa. The occasion was the return to Samoa of Alfred Wyld, a prominent planter and trader, who had been in forced exile in New Zealand. Wyld was one of three white planters who had been removed to New Zealand by the colonial authorities several years earlier. Tamasese himself had been transported to New Zealand to serve a six month jail sentence for refusal to pay taxes and upon his return took up the presidency of the Mau association in the absence of Olaf Nelson, also in exile.

The Mau ("firmly held opinion") movement began among the Samoans during the German colonization of the islands (1900-1914) and continued under the post-war New Zealand mandate until the independence of Western Samoa in 1962. The Mau was a set of institutions the Samoans (including some residents of European and European-Samoan descent) maintained parallel to and in defiance of the attempts by European colonists to establish governing structures that excluded Samoans while subjecting them to taxation, conscription and other forms of exploitation.

The development of large scale agriculture on the islands led to the appearance of managers and merchants (European and Samoan in descent) unwilling to accept colonial control in economic and political matters, and prepared to make common cause with the Samoans in their largely peaceful protests. An epidemic of influenza heedlessly introduced onto the islands in 1918 decimated the Samoan population, but did not diminish the defiance rooted in the villages where the native Samoans resided. In the 1920's a parallel Mau police force formed and sported their own uniforms as they harassed the colonial force.

The governors appointed by the New Zealand government, a succession of military officers who had seen action in World War I, adopted increasingly repressive measures, from imprisonment and exile of Mau leaders to beatings and shootings. Gandhian passive resistance, which the British colonial authorities were facing in South Africa and India, was adopted by the Mau. Tupua Tamasese Lealofi III was shot in the back by police as he addressed the crowd counseling demonstrators to practice peaceful resistance.[257]

The harsh reaction had begun some months before when New Zealand police attempted to arrest a Mau leader who had not paid his taxes. According to witnesses giving evidence at the subsequent coroner's inquest, Tamasese had been warned that if the tax resisters marched in the procession to welcome Wyld they would be arrested. They did march and the police efforts to effect the arrest initiated the melee.[258] The

---

[255] de Montigny (1953: 108)
[256] Naidu (1993: 128)
[257] The last words of the 28 year old Tamasese, engraved in Samoan on his memorial stone in Apia, enjoined his followers not to seek revenge.
[258] New Zealand. Parliament. House of Representatives (1930)

ensuing charges and countercharges led to a police constable being chased down and beaten to death, a machine gun (Lewis gun) atop the police station opening fire in four bursts and rifle fire while the Lewis gun was firing bringing down three of the Samoans who died. The conclusions of coroner J.H. Luxford's report set apart the separate firearms discharges at the time of Tupua's death.

16. Samoans advanced towards the police-station from three directions, and commenced to stone it on the eastern side.

17. Sergeant Waterson employed Lewis-gun fire for its moral effect, and succeeded in turning back the advancing Samoans without causing any casualties. The number of shots fired from the Lewis gun was fewer than forty-seven : they were fired in four bursts.

18. Sergeant Waterson, being an experienced Lewis gunner, was justified in his action.

19. The deaths of High Chief Tamasese and of Tu'ia and Migao were the result of injuries inflicted by rifle-fire delivered immediately before and during the time the Lewis gun was in action.

20. The rifle-fire was delivered by three members of the Police Force from the northern balcony of the police-station, and was directed down Ifi Ifi Road.

21. The evidence does not show that the rifle-fire was necessary, however. In circumstances as then prevailing it is inevitable that some action will be taken which may appear at the time to be justified, but when inquired into subsequently will be found to have been unnecessary.

That is the conclusion to which I come in regard to the rifle-fire which caused the deaths of High Chief Tamasese, Tu'ia, and Migao.

J. H. LUXFORD.

Luxford does allude in the text of his report to the testimony of Samoans who went to the assistance of Tamasese when he fell. Because the machine gun was firing at that time the survivors of the attack were convinced that Tamasese was hit by a bullet from one of its bursts. In later testimony one of these men reconsidered, and did not attribute his wounds to the machine gun.

Luxford concluded that the position of the gun and the trajectory of fire ascertained from testimony made it impossible for the Lewis gun bullets to have struck anyone. A map of Apia harbor included in the report traces the line of fire of the machine gun and rifles. Admitting that it was rifle fire from the police station balcony that killed Tamasese and two of the others, he makes no mention of the ammunition used. He characterizes the circumstances as likely to make it seem the action-the rifle fire-was necessary at the time but upon later examination would be found to be unnecessary. In other words the perspective of the police was clouded by what they conceived of as an imminent attack by club-wielding Samoans. From the viewpoint of the police, the Samoans were the ghazis of the moment.

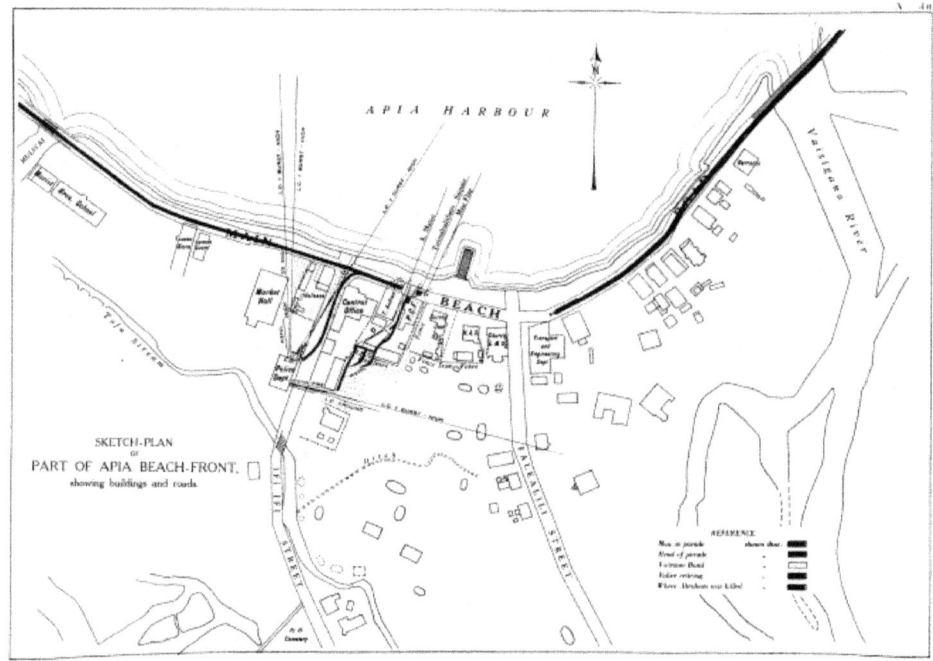

The coroner's report, the most detailed account of the events surrounding Tamasese's death, caused indignation because it seemed to justify the machine gunning of Samoans. Malama Maleisa, a Samoan historian of the islands, quotes a long popular Mau song that places the governor's finger on the trigger of the gun.[259] Other writers either mention the machine gunning, add the coroner's conclusion without comment or state doubts that the machine gunning was the cause of the deaths.[260] Following what came to be named Black Saturday, the Samoan supporters of the Mau took to the villages and suffered violent repression by the police until the election of a Labour government in New Zealand in 1936 brought a change in policy.

The historian of the Mau Michael Field wondered if it mattered to the dying Tamasese whether he had a machine gun bullet or a Lee-Enfield .303 in his body.[261] Field also gives reasons to doubt the coroner's conclusions that the Samoan marchers were not struck by the machine gun bullets.

The Samoan National Assembly was convened by Prince Aloalii Tuimalealiifano at Vaimoso on March 5, 1930 in protest against the violent suppression of the gathering the previous December. Aloalii was a prominent clan leader and a champion swimmer but his assertion that Tamasese and the others were killed in a riot by police shooting dumdums did not enter into considerations by subsequent investigators. The Lewis gun was sufficient to mark the lethality of the shots and obscured the dumdums possibly loaded into

[259] Maleisa (1987: 147)
[260] Wei and Kamel (1998: 107); Thompson (1994: 83); Najita (2006: 82); Belich (2000: 239)
[261] Field (1994;2006)

the rifles. That many of the colonial administrators and police were World War I veterans makes the allegation of dumdum use more credible. Colonial police use of dumdums to suppress protests whatever their character was not unknown though difficult to know until the police bullet policy discussions of later decades.

The Warren Commission report of the official inquiry into the assassination of U.S. President John F. Kennedy was released a year after the November 22, 1963 event. The possibility that one of the bullets that struck the president was a dumdum bullet was raised several times during the testimony of witnesses and experts. The report's conclusion that a lone gunman fired the shots was not compatible with the possibility that just one of the bullets was a dumdum and the others were conventional for the Mannlicher-Carcano rifle.[262] The report did not recognize that dumdums could have been loaded together with conventional bullets in the same firearm. Different kinds of bullets could only come from different guns; the Commission members were certain only one shooter had been present. The report of the U.S. House Select Committee on Assassinations issued in 1979 determined that at least four bullets were fired, three of them from the same gun, but they did not comment on the composition of the bullets.

These contradictory conclusions left open the pathway for reinterpretations of the existing evidence that made the bullet in question out to be frangible, explosive, a hollow point, or a mercury point. Tests subsidized by a television channel established that Kennedy's head would have exploded if a frangible bullet were used.[263] Kennedy's head became the staging area for identifications and suspicions that led back to American or foreign agencies with agendas to be realized by the president's conspicuous death. Intentions of likely sponsors of the assassination were read back into the bullets which were read forward into likely assassins and motives.

Kennedy joined the public figures targeted by the changing bullets with changing origins the more detailed the commentary became. Another head of state was killed with similar dumdums. Israel Prime Minister Yitzhak Rabin was shot on November 4, 1995 by a young law student Yigal Amir as Rabin returned to his car after addressing a peace rally at Kings of Israel Square. Amir was part of a contingent of right wing Israelis who rejected Rabin's signing the Oslo Peace Accords which would grant Palestinians a degree of control over their own population on the West Bank.

Amir bypassed Israel's gun control regulations by claiming a false West Bank residence that called for a personal defense gun. He loaded his 9mm Baretta alternately with conventional rounds and hollow point bullets made by an acquaintance who shared his convictions but not his drive to act.[264] Amir admitted to firing the shots and despite attempting to defend his action on the basis of religion was convicted and sentenced to life imprisonment, a sentence later modified by law to remove any possibility of parole.

The association of a religious zealot with the murder of a prime minister gave rise to conspiracy theories questioning Amir's sole responsibility. A mock assassination was planned for the time of Rabin's speech, and

[262] McKnight (2005: 356)
[263] Chambers (2010: 160)
[264] Horovitz (2012: 206)

at first the security detail thought Amir was firing blank cartridges. The hollow points in this assassination were the most distant from fictional of any assassination.

13. Disappearance

A search of newspapers, journal articles, memoirs and other literature from or pertaining to the period between 1920 and 1948 yields relatively few references to dumdum, expanding, explosive, soft nose or hollow point bullets. Compared to the abundance of writings and images of the 1899-1920 period references to dumdum and expanding bullets have disappeared. The Italian-Ethiopian conflict of the mid-1930's and the Spanish Civil War of the same period generate a slight uptick in references, but the increased warfare in 1940 does not increase the frequency of dumdums discovered, blamed or noted.

In search of an explanation for this change in intensity I examined the expanding bullet instances that were recorded. The exchange of accusations, last rehearsed by the Italians and Ethiopians, was abandoned as a propaganda quarrel. Any party using the word "dumdum" used it as a victim, mostly on the basis of wounds and deaths assumed to be dumdum-caused. The victims might take the assumed dumdum use to be a cause for action. The belief structures of World War I were brutally enforced.

During the British retreat from Dunkirk on May 27, 1940 part of the Royal Norfolk regiment became separated from the main body of troops. Sheltering in a farmhouse in the village of Le Paradis, they engaged the Waffen SS until their ammunition ran out. Their commander ordered them by radio to surrender in the expectation that they would be treated as prisoners of war. The SS conducted them to a wall nearby and machine-gunned 97 of them. Two men survived by faking death, escaped and were sheltered in a nearby farm building. After the war, those men, the French farm owner and another French civilian were witnesses in the 1949 trial of Hauptsturmführer Fritz Knöchlein.

Knöchlein's defense asserted that he was not present during the massacre, that the British used dumdum bullets and opened fire on the SS men while displaying the white flag of truce. Evidence was presented to counter these assertions. Hauptscharführer August Leitl, an ammunition expert with an SS infantry regiment, testified that Knöchlein assigned him to patrol the battlefield and collect the SS dead.[265] He did not find any dumdum bullets. All the corpses had been shot in the head through metal helmets, which caused the bullets to flatten on impact and make large entry wounds. Those determined to find dumdums looked only at the size of the wounds which had the appearance expected of dumdum trauma. Leitl denied that the Royal Norfolks had fired dumdum bullets.

The court ruled that whether or not dumdum bullets were used or the white flag was abused, Knöchlein bore responsibility for ordering the summary killing of prisoners of war. He was convicted and executed by hanging, the only German soldier tried for the Le Paradis massacre. The court concluded that not allowing the Royal Norfolks to defend themselves before an impartial court constituted a war crime.

British troops were once again accused by Germans of using dumdum bullets, a holdover from World War I. This time an ammunition expert present could refute that assertion based on the bullets found and the nature of the wounds. The ideological drive of the heavily indoctrinated Waffen SS must have overcome objections at the time of the massacre, if any were made. Use of the dumdums in past conflicts combined with apparent confirmation in the wounds of dead comrades was sufficient to sentence the soldiers to death.

---

[265] Lee (2006: 57) gives a translation of Leitl's testimony from court transcripts.

The Le Paradis massacre was a throwback only in the search for dumdum bullets to confirm suspicions of their use. Inferring the bullets used from the wounds discovered was a look ahead. During World War II and thereafter the character of the wound was sufficient to prove that expanding or dumdum bullets were in the repertory of the enemy. As in World War I dumdum itself was an absolute: soft nose versus hollow point were not parsed out of the dumdum identification in the wounds, only in the bullets themselves. Each side accused the other of using dumdum bullets but now mostly on the basis of wounds.

The dumdum accusation paired with peace flag abuse was recorded at other times during the war as a reflexive excuse for the killing of prisoners of war by German soldiers. A Roman Catholic schoolteacher who served in the Reich's army on the Western front was transferred to the East where he witnessed the shooting and burial of two Russian prisoners. He was told that one of the prisoners had continued firing while displaying the flag of truce and the other had been firing dumdum bullets.[266]

The discovery of caches of dumdum bullets was sometimes featured in news stories. A photograph that appeared in *Life* magazine in the aftermath of the Allies' Sicily campaign depicts General George Patton holding in his hand for examination by war correspondent Ernie Pyle bullets the caption indicates are dumdums carried by captured Italian troops.[267] The Romanian General Ion Antonescu, allied with Nazi Germany, appears in a photograph published in the Wehrmacht wartime magazine *Signal.*[268] The general inspects a clutch of dumdum bullets the text says the medical officer has removed from injured soldiers. A photograph of bullet cavitation in flesh is suspended beneath the general's bent index finger.

---

[266] Stargardt (2006: Chapter 5)
[267] Aftermath, *Life*, November 15, 1943: 58
[268] *Signal* 1941 n. 16

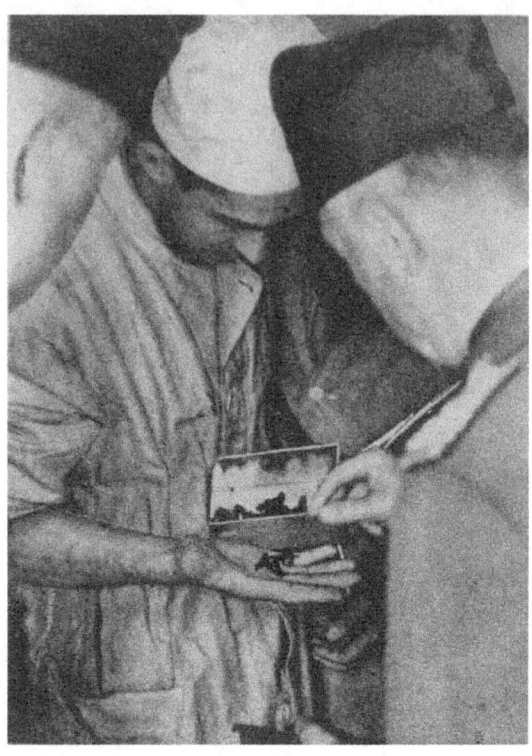

41. General Antonescu views dumdum bullets taken from Romanian soldiers. Fn 265

The surrender of Axis troops at the beginning of the North Africa campaign yielded a "startling bit of evidence of Axis barbarism...eighteen 6.5 mm dum-dum bullets for Italian carbines. The bullets had open lead-nose slits inside their copper jackets."[269] *New York Times* correspondent Drew Middleton assured readers awaiting news of the Allies' success that the dumdum bullets had been issued with regular ammunition but their use was not general and doctors in field hospitals were not finding wounds caused by dumdums. With the allusion to barbarity at the beginning this news item, this story seems to repeat the World War dumdum cache discovery pattern. Middleton instead minimizes the likely impact of the bullets on soldiers in the field. They are mixed in with the standard issue ammunition, perhaps to conceal the occasional dumdum hit amid conventional fire. The story was printed separately from the larger article in English and Australian newspapers,[270] another echo of past dumdums.

One wartime bullet capture that casts into relief the increased emphasis on dumdum wounds took place in 1941 when the Sixth Army of the German Reich came across stores of explosive or dumdum bullets left by the Red Army in Zhytomyr, Ukraine. Oberstabsartzt Gerhart Panning, the senior staff doctor with the

---

[269] Middleton (1943: 4)
[270] For example, Dum-Dum Bullets Prepared for Italian Rifles, *The West Australian* (Perth, West Australia) May 13, 1943

134

Sixth Army, requested permission to test the bullets on live subjects, specifically Jewish prisoners,[271] in order to study their effects on the human body and confirm their illegality.[272] Panning said that he wanted to be ready to treat dumdum wounds in German soldiers. Colonel Paul Blobel, in command of Sk4a, a unit dedicated to implementing the "Final Solution," selected soldiers to carry out the shootings to harden them for further death dealing, but the assignment turned out to be unbearable for members of the firing squad. [273]

After the war Panning's widow initiated "denazification" proceedings on behalf of her husband, who had been a medical forensics investigator into claims of atrocities against the German armed forces.[274] The son of World War I army commander Helmuth von Moltke, James von Moltke, an international law specialist and secret opponent of Hitler's regime, forwarded a document his father had received which describes Panning's positioning the doomed prisoners for the optimum analysis of bullet impact in their bodies. Panning's own published account of the results of his Soviet ammunition investigations artfully evades disclosing the source of his information.[275]

The edition of von Moltke's letters to his wife Freya, published in German in 1988, contains a letter dated December 9, 1941 in which he refers to the discovery of the "Russian dumdum bullets" and that Panning tested them executing Jews "with a shot to the head so and so, and a shot to the breast so and so...These results were scientifically arrived at so that the relevance to international law was unmistakable. This is a high point of bestiality and depravity but one can do nothing about it. I hope one day that it will become possible to bring the involved officers and Panning before a Law."[276]

Panning died of tuberculosis in 1944 and early the following year von Moltke was tried and executed for conspiring against the state. I translated his "ein Gericht" as "a Law." A devout Christian, von Moltke believed that if human law did not punish the culprits a higher one ultimately would.

The Nazis evidently had supplies of dumdum bullets not captured from other armed forces. On February 2, 1943 Jakub Goldberg wrote that he saw the corpses of Jews in the street killed with a shot to the head.[277] Only the neck or the lower jaw remained, making it impossible for relatives to identify the dead. This anonymizing devastation can only have been produced by dumdum bullets.

Martin Pegler found in his interviews with World War I soldiers that the conviction dumdum bullets were the cause of explosive injuries persisted for decades after the war. [278] A head abruptly vacated of its contents or a long bone reduced to slivers formed images that persisted long in memory with the dumdum causal association. Yet the .303 caliber service bullet projected at a high velocity was capable of this damage without alterations to the jacket. Men entering the field of battle for the first time were not aware of this. The dumdum complex was imprinted despite the lack of confirmation.

---

[271] Several sources indicate "Soviet" prisoners, but "Soviet" and "Jewish" are not incompatible.
[272] Lower (2005: 241)
[273] Lower (n.d.)
[274] Preuss and Madea (2009)
[275] Panning (1942) known to me only by descriptions in other sources.
[276] von Moltke (1988: 286)
[277] Gessen (2004: 163)
[278] Pegler (2011: 90-92)

Of course the repeated dramatic publicity of the discovery of dumdum caches and dumdum-filled clips during World War I seemed to provide confirmation that deliberate modification of bullets was the practice of the cruel enemy. Testimony to the contrary, that bullets made or modified to expand were rarely found in captured clips, did not enter into the mainstream.[279] As with the dumdum accusations that doomed the Royal Norfolks and the Russian soldiers at the hands of the Germans, sudden death of a comrade or relative from a sniper's bullet has to have been the result of the immemorial dumdum. The twist of barbarity in the bullet's reputation compensated for the accident or skill behind a conventional bullet striking exactly in the right place.

Wounds caused by snipers were the most widespread evidence for dumdum bullets during World War II. The dumdum complex echoing from the previous war personalized a war that included the largest number of military and civilian casualties up to that time. The shooter was only known as an enemy force but the victim was someone nearby. The proximate wound was the perfect object just as the bullet itself, grown obscure and disenfranchised, was the perfect subject.

Prior to the D-Day invasion an American army military intelligence manual observed that "enemy snipers have used "dum-dum" ammunition, which made it more difficult to locate the spot from which the shot was fired but easier for the enemy to observe a hit."[280] Presumably anyone hit by a dumdum bullet would fall immediately, not revealing the direction it came from but giving the sniper a sign that the bullet had hit its mark decisively. The fighting conditions in Normandy, *bocage* farm fields surrounded by hedgerows, a difficult terrain to cross and an easy one to hide in, would have lent itself to this type of dumdum use. Both sides were said to have refrained from making dumdums in the field because the other would have retaliated in kind.[281]

Confirming that dumdums were in use by snipers or riflemen required examination of the wounds and finding the bullets. The only wounds attested as the product of dumdum shots among all the wounds received in the fighting were attributed to wooden bullets. Major Legere was wounded in the thigh as he led a group of men in an assault on the village of Pouppeville. The gaping severity of the wound caused the men to conclude that the missile was a wooden bullet, verified later by finding casings mounted with purple-dyed wooden bullet noses among the ammunition of dead German soldiers.[282]

Since there were no shaved or hollow-nosed dumdums discovered, the wooden bullets had to serve the purpose of explaining unusually severe sniper wounds. They were training bullets that German soldiers carried and sometimes fired in battle when they ran out of other ammunition. They were made of balsa wood and not likely to do more than leave a mark.[283]

[279] Hesketh-Pritchard (1920: 29-30)
[280] U.S.War Department. (1943: 37)
[281] Beevor (2010)
[282] Koskimaki (2006: 161); McManus (2004: 229). Marre (1916: 29) asserted that the Germans used wooden bullets during World War I.
[283] Ambrose (1994: 140); Pogue (2001: 122)

A recollection of the push toward Kokumbona on Guadalcanal during November, 1942 included a duel between an American rifleman and a Japanese sniper.[284]

> Cox, one of Boyd's riflemen, poked up from a log at the gully's mouth and tried to get the sniper somewhere up the gully. The sniper caught him first though and dropped him with a dum-dum slug through his right collar bone and shoulder.

In another Guadalcanal episode a Marine was between two men on one side hit by bullets from a .25 caliber submachine gun, and another man on the other side hit in the forehead with a dumdum shell spreading brains and blood over him and lodging a piece of skull in his shoulder.[285] Unlike the wounds of the other two men, who were felled by the .25 caliber bullets, this explosive wound must have been caused by a dumdum.

Wounds like this reinforced the assumption that Japanese snipers were shooting dumdums while others nearby with less eruptive wounds were shot with different bullets. Yet there were no reports of dumdum bullets of any description found in caches or among the ammunition of captured soldiers. An intelligence report issued later in the war gave specifications of the "Japanese 6.5mm dumdum," which was the Arisaka bullet of the Russo-Japanese war, though the report didn't say so.[286] The formerly humane bullets were Japanese dumdums during World War II, in keeping with the brutalized image of the enemy.

The Japanese extended their empire through the conquest of existing colonies of European powers and America in the Pacific and East Asia. They also exploited the indigenous peoples but posing as liberators and formed administrations composed of elites eager for independence from their foreign masters. With the surrender of Japan on September 2, 1945 the colonial powers reasserted their control and were met with revolutions in some places encouraged and supported by the departing Japanese. That support included the release of firearms and ammunition to militias and other armed bodies resisting reoccupation.

British forces in the southeast Pacific attempted to secure the Netherlands East Indies for the Dutch who had operated profitable enterprises on the islands for centuries before the Japanese conquest. The resistance movement in what would eventually become Indonesia was a postwar challenge. The Dutch government had offered in 1942 to recognize the independence of the colony as part of the Dutch Commonwealth. The rebels sought full autonomy and struggled with British and Dutch forces to capture territory and destroy the institutions of economic and political oppression, for instance rubber plantations.

The ensuing warfare involved military, paramilitary and civilians. The dumdum complex was staged to a greater extent than during the war itself. Through the communication media available to them, the opposing sides accused each other of using the bullets. During the fighting for possession of the cities of Surabaya and Batavia on Java, in November, 1945, a British spokesman reported by an Australian newspaper said that Gurkhas had found dumdum bullets at an outpost where Dutch hostages were being held, and that they had been fired in Surabaya and Batavia.[287] Dutch soldiers fighting with the British for control of Batavia were shot "apparently by dum-dum bullets."[288] In the Medan area of Sumatra, where the Indonesians made land mines

---

[284] Rogal (2010: 79)
[285] Lino (1944)
[286] U.S. Army. C.B.I. Ordnance Technical Intelligence Report. No. 39-1. February, 1944.
[287] Crisis Deepens as Dutch Ban Talks with Soekarno, *The Canberra Times,* November 5, 1945: 1
[288] British Forces in Control of Batavia, *The Canberra Times,* October 17, 1945: 1

from Japanese airplane bombs, the heavy Dutch casualties were attributed to the Indonesians resorting to dumdums.[289]

The finding of bullets and the high death rate were accompanied by accounts of wounds that could only have been caused by dumdum bullets. In Mrs. van Hannem's recollection of the evacuation of Dutch women and children from Surabaya under fire by Indonesian insurgents a woman's bleeding from a bullet that struck an artery was controlled only for her leg to be "completely torn off" by a dumdum bullet.[290] The bullet that blew off the leg of a woman who had survived the previous bullet had to be a dumdum.

Within their own media Indonesian independence fighters leveled accusations against the Dutch which were sometimes reported more widely. Broadcasts of Jogjakarta radio, established by a sultan who supported the Indonesian independence movement, aired accounts of Dutch attacks on Indonesian communities in the course of the war, among many other actions an assault on Malang in Western Java with dumdum bullets.[291]

The word "dumdum" with its onomatopoeic battle sound and associations of violent mutilation by foreign attackers passed directly into many languages. The Dum Dum locale in India became the site of an important military airbase (and eventually Calcutta's international airport). In South Asia at least dumdum bullets and airplanes were linked without any distinct basis in military technology. Strategic bombing of Indonesian sites by the Dutch encouraged the association.

During the late 1950's, after Indonesia had been recognized as an independent nation, a woman named Sinek beru Karo recorded a sung narrative in the Karo language of the 1947 evacuation of an area of highland Sumatra under pressure from Dutch forces. The Dutch were attempting to rout the rebels after the British had turned control of the Medan area over to the Dutch the previous year. Sinek shaped her song according to the katoneng-katoneng form, traditionally composed for recital during funerary rites, and a means of recalling communal events during the independence struggle. Cassette recording technology preserved and disseminated among the Karo performances of this and other songs of the war.

Stanzas of the song, as translated into English by Mary Margaret Steedly from a transcription of a recording, recall the mechanized violence of the Dutch onslaught upon the displaced villagers and townspeople.[292]

> They came by land
> Bringing all their tanks
> Shelling first with mortars
> And they came from above with airplanes
> Firing down constantly
> With lots of big dumdum bullets.

"Dumdum" was one of the words that remained the same from Dutch to Karo to English: Sinek believed that this named the bullets fired by the Dutch fliers as they suddenly swooped down on the columns of fleeing refugees. Trying to evade the death and suffering wrought by the ground assault, the Karo were beset from the air with an inescapable force.

---

[289] Dum-dum Bullets Used by Indonesians, *The Canberra Times* December 4, 1946: 1; Jenkins (1946)
[290] McMillan (2005: 45)
[291] Indonesians Urged to Give Support to Army, *The Canberra Times* October 17, 1945: 1
[292] Steedly (2000: 827)

Steedly met Sinek during her fieldwork in Sumatra and learned about the events of her life from her becoming a singer against the disapproval of her father (women being relegated to the status of perpetual minors in Karo society) to the dislocations of the evacuation period, which she crafted into a narrative of wide appeal. In her song, however, she omitted that the group she was fleeing with were not entirely Karo but were political prisoners being relocated under her father's leadership. She also left out the attacks upon her group by Indonesian nationalist militia and other acts of violence by armed bands. To be a memory artist for her Karo clanspeople she had to portray a common emergency precipitated by unseen outsiders. The Dutch woman refugee whose leg was torn away by an Indonesian bullet was part of another group's story.

Dumdum bullets disappeared from public discourse during World War II, but they reappeared suddenly in the discourses of groups in revolt, under attack, subject to oppression by militarily superior adversaries. The word, the bullet and the sudden tearing of flesh come from the outside. Unlike the exchanges of accusations during the period when the head of state of one country could send a telegram to another or to a League of Nations protesting the use of arms declared illegal by international conventions, the conflicts of the post-war were not conducted between parties with equal access to the means of broadcast. This did not prevent any group from receiving knowledge of expanding bullets, the technology, the terminology and the victimology.

The Arab-Jewish conflict fired up by Zionist settlement in Palestine from early in the century and reaching a fever pitch with the formation of the state of Israel in 1948 went through an expanding bullet evolution of its own. The word "dumdum" is used in European and American newspaper accounts of attacks from the 1930's onward, but not in Arabic language newspapers until much later. Jewish settlers and British soldiers are shot at by Arabs or by assailants assumed to be Arabs and the victims are hospitalized or autopsied, the bullet type then determined.

In April, 1931 the watchman at a Zionist colony and agricultural school near Haifa was shot in the foot with dumdum bullets when he challenged "armed thieves" attempting to enter the colony.[293] Mandate police followed the assailants to a nearby Arab village and arrested 50 suspects. Arabic language newspapers deplored this and expressed sympathy for the relatives of three young Jews shot to death near Haifa. As the 1930's progressed and dumdum accusations were passed about, between the Italian and Ethiopian governments, against the nationalists in the Spanish Civil War, against Bolivians by Paraguayans, the charges of Arab dumdum use multiplied.

When Arabs fired on a bus carrying Jewish passengers and British troops were called in, the "miniature battle" that resulted left Jewish and British casualties found to be wounded with dumdum bullets when examined in hospitals.[294] Jewish settlers working on a road project near Jerusalem were fired on with dumdums and instantly killed by an "Arab gang." Pursuers found the cartridges of their bullets.[295] The map of the Arab campaign in Palestine for 1936 includes other instance of Arab dumdum use.[296] In a reflexive move,

---

[293] Crime Wave Follows Slaying in Palestine, *The New York Times* April 10, 1931
[294] Arabs and Troops Fight in Palestine, *The New York Times* May 22, 1936
[295] Arabs in Ambush Murder 5 Zionists, *The New York Times* November 10, 1937
[296] Gilbert (2005: 21)

the British War Office issued a communiqué denying that British troops were using dumdum bullets in Palestine.[297]

The anti-Jewish riots that directly preceded and accompanied the formation of the state of Israel took place wherever the two groups were in contact with each other. A Jewish girl and man shot to death during the December, 1947 outburst in Aden (South Arabia) were found in postmortem to have been the victims of dumdum bullets.[298] They were the same caliber bullets as those issued to the Aden Protectorate Levies, the peace-keeping force organized by the British administrators of the protectorate. Making conventional bullets into dumdums was a skill known to both Arabs and Jews, but suspicion pointed to the Arab rioters.

In instances like this the beginnings of a transition are visible. The dumdum making skills are mutual. During the 1930's there were reports of Jewish attacks upon Arabs in Palestine but there were no dumdums mentioned. After the war the order is reversed. The Jewish settlers have become Israelis (as have some Palestinians). The Israeli police and military defend themselves from charges that they use expanding bullets to disable Palestinian protesters, whose missiles usually are stones, hand-held signs (sometimes in English for media exposure) and words.

Bands of armed men attacking settlers meet a response from the settlers themselves and from Israeli armed forces. Eruptions of discontent over incidents arising from mistreatment of individuals and land development threats to local communities are treated as threats to the civic order of the state. The same conditions of revolt and repression that led to (the accusation of) police use of expanding bullets (or bullets that seemed to be) during the 20th century from Ireland to America to China also existed in Israel/Palestine. The Nazis, in their attempts to police their conquered territories set a standard that others unknowingly followed.

The Arab-Israeli encounters have been magnified both by the increasing breadth of coverage of events (photos, film, video, radio, internet) and by the international representation of both parties. The us-them effect that has long given a directionality to dumdums (us-Jewish settlers/them-Arabs) was reversed. Instead of the charge-countercharge formation of World War 1, where each side took up each position in alternation, the Israeli-Palestinian conflict is formed as a proposed us-Palestinians accusing a resistant them-Israelis.

Opinion and alliance can reverse the two and the Arabs can be made out to be the aggressors. Arabs or any one group identified as Arab are not said to shoot Israelis with dumdums as the pre-independence Arab attackers were reported to do. As with their own reputation the British long labored to quash, the dumdums in Israel are asserted to be fired from the weapons of the Israeli police and armed forces wherever dumdums are in evidence.

In 2002 the Institute of Community and Public Health at Birzeit University (Ramallah, West Bank) published findings of a survey of the wounded during the most recent intifada, an anti-Israel uprising by Arabs, which began on September 28, 2000 and was ongoing at the time of the survey.[299] The uprising was stirred by Israel Defense Forces separating and controlling Palestinian territory, the construction of new

---

[297] Dum-Dum Bullets Not Used by British Troops, *Nottingham Evening Post* June 8, 1936: 7
[298] Edwards (2004: 96)
[299]Halileh, et al (2002); Ferriman (2002)

settlements on the West Bank, and the lag in implementing the 1993 Oslo agreement that ended the previous intifada. The Birzeit survey examined cases of non-lethal injury during the first 94 days of the conflict, between September and December, 2000.

A sample of 6071 injured between September and December, 2000 contained 25 percent schoolchildren and 60 percent aged 18-34. The rate of permanent disability was estimated to fall between 15 percent and 39 percent of those wounded. The burden of disability was disproportionately borne by children and young adults. Live bullets, rubber bullets with a metal core and inhalation of gas from canisters were the agents of injury. The high rate of disability from bullet wounds was due to the tumbling and fragmenting bullets fired from U.S.-made M-16 field rifles. On meeting skin the bullets turn and spread their kinetic energy end to end tearing flesh and pulverizing bone, comparable to dumdum injuries. Many of the wounded children were hit in the legs as they ran.

The Israeli military (IDF) and police had been armed with M16s and the 5.56 mm bullets they accept since before the first intifada in 1987. During the late 1960's to early 1970's they acquired inexpensive M-16s to replace Soviet assault rifles captured from Arabs, the Kalashnikov AK-47. A Kalashnikov derivative, the Galil series of assault rifles, was manufactured in Israel during from the early 1970's but never was adopted wholesale by the IDF because of the large number of M-16s arriving at the time of the Yom Kippur War in 1973 and thereafter. Both M-16 and Galil exist in 5.56mm and 7.62mm versions, the 5.56mm favored for the lightness of the ammunition load.

Besides tumbling and fragmenting, 5.56 mm bullets throw out shrapnel on striking stone close to the victims, some of whom were standing near walls and other stonework. 2005 May Day demonstrators in Bil'in (West Bank) fired upon by Israeli police gave evidence of "live exploding (dumdum) bullets" in the X-rays of their shrapnel wounds posted online.[300] The dumdum nomenclature gives an accent of international disapproval to the act of wounding. Illustrations of dumdum violence are specific and local; attempts to indict and defend the Israelis have a far-flung character.

At the time of that first intifada when Palestinians faced massed Israeli response to organized demonstrations, the dumdum charge was advanced by Arab media and foreign observers and countered by Israeli officials. The Israel Foreign Ministry sent a telegram to the Foreign Office in London complaining that Neil Kinnock, Member of Parliament and Labour Party leader, had made an unfounded statement to the press that the Israelis were putting down the protests with dumdum bullets.[301] Kinnock privately agreed that the statement was unsupported but never made a public retraction.

A 90 minute documentary film entitled *Days of Rage: The Young Palestinians* produced by the Public Broadcasting System (PBS) in the United States contains interviews with Palestinians who attribute their wounds to Israeli dumdum bullets. Eric Goldstein of Mideast Watch wrote an article for *The Philadelphia Inquirer* (September 6, 1989) decrying the distorted picture of the still ongoing conflict given by the film. Goldstein cited the opinion of Palestinian doctors who had examined injured civilians and found no

---

[300] Protestors Hospitalized After Anti-War Demonstrations in Bil'in, *Online Intifada* May 2, 2007
[301] Schleifer (2006: 220n53). Telegram of March 10, 1988.

evidence of dumdum injuries.[302] The producer of the film denied that the film was financed by Arab interests, but the **PBS** executive who had acquired the film resigned her position.[303] In December, 1989 **PBS** announced that a film entitled *A Search for Solid Ground: The Intifada Through Israeli Eyes* would be broadcast in January.

The disabling injuries caused by M-16 bullets in a portion of those hit were classified by dumdum rhetoric. The Palestinians remained the object of the Israeli subject. Both sides attempted to qualify that relationship further rather than deny it entirely. Expanding bullets continue to be a simple assumption. Now the plethora of information about attacks, attackers and victims, the variety of media and the diversity of viewpoints represented challenges the dumdum charge on all sides. The bullets disappear only to reappear again in a different shape. M-16 bullets become dumdums become disabling bullets that can be received many times yet survived. The Palestinians are positioned to receive the bullets and declare what they are; the Israelis are positioned to declare what they are not, but not what they are.

A 25-year old Palestinian man, Ibrahim Bornat, taking part in a demonstration against a wall constructed by the "Israel Occupation Forces" in his village of Bil'in was shot three times in the thigh with what were described by the piece on him as dumdum bullets.[304] Ibrahim told the reporters that he had been shot 77 times already. He already bore a scar from a gas canister that had grazed his forehead. The bullets had torn an artery in his leg causing major blood loss and a need rapidly to find donors of his uncommon AB blood type. Damage to bone and muscle required surgery and the prognosis was uncertain. Ibrahim would require considerable rehabilitation to enable him to walk. He complained that the Palestinian Authority, who were supposed to pay the medical expenses of injured Palestinian activists had not been forthcoming, and private contributions had to be solicited. A newly constructed hospital nearby only treated patients with insurance.

Ibrahim's elder brother Rani had been shot in the neck with a dumdum on September 28, 2000 and was completely paralyzed. The two brothers are exemplars of the toll of Israeli bullets on Palestinian families, and the difficulty obtaining adequate care. In cases like that of Ibrahim emergency care made the difference between life and death, between mobility and disability. The only dumdums among the total of 80 bullets Ibrahim counted, were the ones that confined him to a hospital bed.

Prior to his thigh injury Ibrahim had collected metal core rubber bullets and other pieces of Israeli ammunition. He had containers of the rubber bullets found in the street as well as casings and live ammunition. Most of the 80 bullets that struck him were rubber, which do not enter the body. The last three were live ammunition, most likely M-16 bullets, though he does not say that they were recovered from his wounds. An outline of a dove made of bullets attached to a white board was one of Ibrahim's contributions to an exhibition, *From the Scent of Bil'in's Wall* that opened the month before his debilitating injury.[305] Ibrahim resembles Sinek, making art from what has been impersonally propelled into him.

Ibrahim knew dumdum bullets by their disappearance into the wounds and debility they caused him, his brother and other victims of the IOF. Other bullets were recovered whole and were displayed. This

---

[302] Lahav (1993: 194n57)

[303] Gerard (1989a) and (1989b). Gerard reported on the *Days of Rage* controversy in a series of articles,

[304] Awad and Jamjoum (2008)

[305] Jamjoum (2008)

disappearance had begun with World War II and continued into the conflicts between states and repressed groups after the war. The disappearance was from the displays of captured dumdums, the mushroomed bullets removed from bodies, the X-ray photographs of bullets pulverizing bone are artifacts of the previous era. In the early 21st century even the X-ray images do not include the bullet.

Verses by the Kurdish poet and memory artist Ahmed Arif (1927-1991) include the refrain[306]

> Kinsman, write my story as it is
> Or they might think it a fable
> These are not rosy nipple
> But a dumdum bullet
> Shattered in my mouth

The phrase translated "dumdum bullet" is "domdom kurşunu" in the Turkish original, the name of the bullet having been adapted to Turkish phonetics. The dumdum bullet reappears in the place of a young man's erotic dream.

The speaker of these lines, as any Turkish Kurd, and many Turks would know, is one of the 33 Kurds arrested for smuggling goods across the Iraq-Turkey border and executed in July, 1943. The Turkish general who ordered the executions was tried and convicted of murder in 1950, but a gendarme barracks built in the village of Özalpi where the incident occurred was named after him.[307] Arif's poem recalls names and places of the event, relations among the executed, the Kurdish land in Eastern Turkey near Lake Van, and the rumor that dumdum bullets were used. A further reminder was the execution of 35 Kurds for smuggling in 2011. The dumdum bullet pins down the sudden injustice of the deaths.

The Kurds are a stateless people alien in three states: the police and military of Iran and Iraq have aimed bullets at them as well, some on the right trajectory to be dumdum bullets. The Vanni of northern Sri Lanka are caught between a state, Sri Lanka controlled by the Sinhalese majority, and the Tamil rebels attempting to establish a separate polity in their region. As elsewhere the increasing availability of arms has made the conflict into an exchange of lethal projectiles, some of which fall within the dumdum verbal category.

The Vanni, living in the village of Mullivakal, targeted by gunshots, sought shelter in bunkers.[308] Outside the bunkers the trees were riddled with bullets, "some described as 'dumdum' a type of bullet." A "dum" sound was heard and an elderly lady thought her leg was broken. "Later we realized that a channam (round or bullet) had struck her leg and then exploded again within." Another type of missile called a cannon was not heard until after it exploded.

For this group of Vanni the dumdum name was attached to the audible sound of an internally exploding bullet. The bullet had vanished to the extent of seeming to be a broken leg until the sound was heard.

---

[306] From the translation of Ahmed Arif's Otozüç Kurşun (Thirty-three Bullets) by Murat Nemet-Nejat at www.turkishclass.com/poem_167
[307] Yeğen (2011: 77)
[308] Sonasundaram (2010)

14. Evolution

Expanding bullets changed as they were aimed in the same direction toward outsiders, who themselves imbibed the notion of the bullets together with the bullets themselves. The expanding bullets had originated in the attempt to make a bullet able to penetrate hostile alien bodies. The design underwent an evolution allowing that intent to be concealed within a victim's changing body that ever sharpened and obscured the intent. The disappearance of the bullets evolved as well.

An end point of this evolution was reached in the manifesto and actions of Anders Behring Breivik. Breivik is under containment in a Norwegian prison for at least 21 years. He was convicted of having detonated an explosion that killed 8 people in Oslo, then shot to death 69 Labour Party youth leaders at a retreat on Utøya Island. The July 22, 2011 killings were the beginning of a revolution that would expel Muslims from Europe and restore the primordial patriarchy that Breivik believed once was the common government of European peoples. His 1,518 page manifesto would stir the Knights Templar to rise and end the multiculturalism that was corroding the spirit of Europe.

A section of the manifesto (3.31) is devoted to "Equipment-weapons/ammo/armour." Here he advises his warriors on the types of ammunition to load into clips of the guns also listed.[309]

Ammo (clip administration)

Assault rifle ammo: 1 armour penetration round, 1 dumdum round (ratio 1/1)
Handgun ammo: 1 armour penetration round, 1 dumdum round (ratio 1/1)

Breivik expects the armor penetrating rounds to precede the dumdums one after another in automatic fire. There follow pages of other sections on how to prepare the armor, how to acquire the guns, explosives, chemical, biological and radiative warfare. In the chemical warfare section he relates how he obtained 50 ml of 99% pure nicotine from a Chinese chemical supplier. Placing this fast-acting poison in a recess in the nose of a dumdum bullet added to the assurance that the round would be lethal.

Breivik expected to combine explosive, expanding and poison bullets in the same assault. Analysis of the ammunition he had with him when captured, and of the wounds of his victims makes it clear that he achieved at least one of these aims. Dr. Colin Poole of the hospital that received 17 of those shot told the Associated Press: "These bullets more or less exploded inside the body. It's caused us all kinds of problems dealing with the wounds they caused, with very strange trajectories."[310] Reports that he poisoned the bullets were based on the manifesto, not on concrete evidence.[311] A newspaper investigation of Breivik's eBay purchases uncovered that he had ordered from a British firm a drill press vice which he wrote in the

---

[309] Breivik (2011: 860)

[310] Shooter Used Special Bullets for Maximum Damage: Surgeon, *Sydney Morning Herald* July 25, 2011

[311] Breivik Injected His Dum-Dum Bullets to Make Them Deadlier, *Mail Online* July 25, 2011. Helen Pidd, a journalist who attended Breivik's trial, wrote in her blog that Breivik said he abandoned the use of poison because he had "no back up."

manifesto would hold the cartridge in place while he drilled a pocket in its lead core.[312] A shadow of that World War I cartoon (Figure 31).

The surgeon's statement was the only evidence that Breivik fired dumdum bullets at his victims; there is no evidence he fired armor penetrating rounds. According to the manifesto they were to kill armored police officers he expected to assault or resist. He was dressed in a fake police uniform and a bullet-proof vest when he was captured. He did nothing to provoke the approaching police to open fire. He intended to survive the massacre to distribute his manifesto and preach his doctrine, which since his detention has attracted a scant following and a few expressions of support.

Breivik wrote in the manifesto that he was forced to take knitting and sewing courses in his primary school in Norway. That effort to "feminize" him and other males failed because he used the skill he gained to design, assemble and evaluate body armor. An illustration in the text juxtaposes an armored knight with a contemporary body-armored figure. The body armor is the resistance of the defender and the assailant rolled into one.

Breivik evidently did use multi-shot magazines in his crime, likely the 10x30 round .223 caliber magazines he says he purchased from a small American supplier. The rifle he holds in a photograph in the manifesto is a Ruger Mini-14 fitted with Eotech and a 3x magnifier.[313] The banana magazine is visible. He had dropped the rifle by the time the police arrived and had been shooting with the Glock handgun he also carried.

The bullets he fired have not in themselves been described in press releases. They would have to be 5.56x45mm (.223) bullets for the Ruger and Glock. They are the standard NATO bullets obtainable from a Norwegian supplier. Hollow nose 5.56x45mm's are used in small game hunting and Breivik could have purchased them using hunting as an excuse, as he did to purchase the Ruger. He did not pursue his attempt to acquire a 7.62x39mm Ruger Mini-30 because trying to obtain a license for a second rifle would have brought inquiries.

The 5.56mm bullet does fragment and yaw at high velocity. Its effectiveness in killing a living target within its optimum range depends upon the training of the shooter and the accuracy of target acquisition. Breivik trained himself at a shooting range (since closed) and provided his Ruger with a scope the better to plant the bullets in susceptible body locations. Breivik quite likely purchased the bullets in hollow point versions for hunting. Only the surgeons' statements to the press and Breivik's own writings support dumdum use.

The "necessary cruelty" of the attack on "cultural Marxists" and "multiculturalists" called for severe means. Breivik may or may not have fired poisoned hollow point bullets. He certainly conveyed the intention to do the lethal damage he believed they would cause. The 5.56x45mm bullets, hollow point and poisoned or not, did cause the injuries the surgeons struggled to mend, killing 69 of the 110 shot.

The shootings replicated the imagined conditions leading to the creation of dumdum bullets in the first place. Those dumdums were specially made to stop foreigners, mainly Muslims, from charging the barricades of the defenders of civilization. The only Muslims among the youth Breivik was shooting at were Chechen men, who were familiar with armed assaults and planned to knock him over with stones until they

---

[312] Barrett (2011)
[313] Farago (2011)

145

decided not to take the risk of attracting his attention. The rapid-fire rifle with magazine, the presumed dumdums, the fake police uniform and genuine body armor were the end point of evolution: Europe's self-appointed knight facing numbers of unwarned and unarmed youth on an island and not the formidable adversaries of the overseas colonies.

Breivik's killings and their fantasy framework were one end to the evolution of the expanding bullet repeatedly reached. The fantasy was little different from the appropriation of victim status by defensive colonialists manning the lines against advancing natives: now they are defending their home country, or rather home continent, against the formerly colonized, now refugees. Breivik is an action hero marching forth to strike against the invaders. He is a pompous coward attacking unarmed people isolated on an island with the entire tradition of dumdum certainty.

Breivik's stated intent and deed manifested wish to cause pain and death to the victims, an inversion of the humane bullet pretense of earlier years. Breivik's apologetics for his action, I regret I must do this, is reminiscent of the murderer who tells his victims that "the best I can do" is flip a coin to decide whether I shoot you or not. But that murderer did not set his ammunition to torture the target. The gratuitous and domineering injection of pain evolved within the technology of the bullets toward outcomes without the admission of intended torment.

Shortly after dumdum bullets, their technique of production and their name began to spread another type of expanding bullet with similar pretense to stopping power was in the process of formation. The listing of British patents for 1897-1900 includes several variations on this maturing theme.

Neville Bertie-Clay, the reputed originator of the dumdum, entered a patent for a compound bullet made of two metals, a brass casing (A) and a lead core (C). [314] A hole at each end of the casing allowed the lead made fluid by the heat of the explosion to flow out of the casing, forming a mushroom shape (Fig. 7) "thus enlarging the incision caused by the bullet."

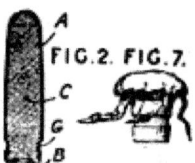

42. Bertie-Clay compound bullet patent diagram. Fn 311

This was an attempt to realize caliber expansion by releasing the lead from the casing to grow to a greater diameter as it entered the body. To prevent premature separation of the core and casing and consequent lead fouling of the rifle barrel, Bertie-Clay ran a ringed groove (G) and coned the base of the casing (B) inward.

---

[314] Patent No. 17, 996 granted July 31, 1897. Great Britain. Patent Office (1902: Class 9, 13).British patents were numbered consecutively starting with the beginning of each year.

Perforating the casing rather than shaving back the jacket as with dumdums was a tacit admission that the lead exposure at front had to be controlled. Other bullet designs in this same set of patents opened the combination bullet jacket at the nose and bent the metal uniformly inward (H.S. Maxim, 14,717, 1899). Both approaches also attempted to keep the lead out of contact with the gun bore but allowed it to spread upon impact with the surface of the target. These bullet types did not enter into production but they were models for later attempts to create foul-free expanding combination bullets.

Another reworking of bullet cross-sections projected differently shaped recesses in the top and the bottom of the unjacketed lead bullet. This was Patent No. 14,754 (1897) by T.W. Webley. As the patent illustration shows, the walls of the bullet were meant to fold back on impact. The base recess was to expand and fill up the gun barrel adding to the propulsion from the expanding gases of the explosion.

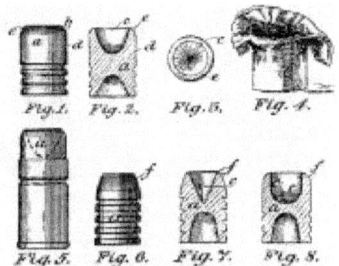

43. T.W. Webley "manstopper" bullet patent diagram G.B. 1897: 14, 754

Webley had combined the open-nose mushrooming bullet with the Minié ball of the mid-19th century.

The Minié ball was a lead bullet with a conical top, a deep recess in the base and ridged grooves running about the perimeter like the Webley bullet. Between the Crimean War and the American Civil War, its main theatres of operation, the Minié ball's recess underwent alterations in shape from triangle to parabola. The bullet arrived in a paper cartridge with a quantity of black powder. The rifleman tore the cartridge open, poured the powder into the rifle, rammed the ball down so its base filled with powder in the barrel, then fired the gun. The heat of the powder ignition caused the base to expand and the ridges of the grooves to engage with the rifling of the barrel as the bullet was impelled forward by the gases with an aerodynamically stabilizing spin. The increased accuracy and force of the bullets led to shattered bones and ruptured organs: warfare had become more deadly at a distance.

Webley aimed to add the mushrooming open nose to that projectile force. The outside walls of the front were slightly chamfered to give the bullet a frontally tapered profile and increase the air-dividing drive of the bullet, to recover the aerodynamic advantage of the conical shape of the ball. This structure also made it a wad-cutter: it punched a hole in the target's skin just before mushrooming out. "Expansion commences immediately," the company's publicity screed continues,"and after the bullet has travelled six inches, it produces a jagged hole from three to four inches in diameter. A wound such as this would doubtless be sufficient to instantly finish even a fanatic."[315]

Figures 2, 7 and 8 in the patent illustration above are variant shapes of nose and base recesses derived from Minié ball history. A number of service bullets made in Europe and America copied the Minié ball's recess with modified nose but Webley's expanding bullet was the first to expand both ends, the base in an orderly engaged manner and the front to fold out, a dumdum less likely to foul the barrel.

In 1897 the family gunmaking firm Webley headed merged with two other riflemakers to form The Webley and Scott Revolver and Arms Company, Limited of Birmingham. At the annual meeting in February of the following year, company officers reported that sales expectations for the "manstopper" bullet had not been met.[316] The company relied on contracts to supply the armed services and police with arms and ammunition, but promotion had not been sufficiently "forward." The competing Mauser, the business report continued, was not suitable for close range combat because "the small projectiles went straight through the advancing man, whereas the man-stopping bullet expanded, incapacitating the advancing man." As a means of halting the charging enemy, the Webley bullet held the same putative advantage over the Mauser as the dumdum did over its fully jacketed original.

The bullet was loaded into hollow faced cartridge case types in at least 3 different calibers.

.380" rev   .450" rev   .455" Mk II short case   .455" Mk I long case

44. Webley "manstopper" bullet loaded in cartridges

The open front allowed the bullet to exit the shell aligned with the bore. Loaded with cordite powder, the bullet was adopted as the British army service bullet under the name Webley Cordite Mark III. Despite the company's great expectations, the bullet had only a two year career in that capacity.

The term "manstopper" was coined by the company and implanted in public consciousness through newspaper articles extolling the bullet's stopping power compared to the existing service bullet. H.J. Ogden of the National Reform Union and Alfred Marks brought these newspaper pieces into evidence for their denunciation of the bullets.[317] Marks contrasted Webley publicity for the "expansive bullets" with Lord Roberts's widely reported indignation at the "explosive bullets" allegedly being fired at British troops in South Africa by the rebellious Boers.

---

[315] Quoted by Marks (1902: 633)
[316] Report on the meeting in *The Statist* 43 (1898): 305-06.
[317] Ogden (1901: 146); Marks (1902: 633-34).

Without acknowledging these comments and international pressure embodied in the Hague Convention, the British government withdrew contracts for the Webley Cordite Mark III, ordered existing stocks to be used for target practice only, and called for overseas stocks to be replaced by the Webley Cordite Mark II, a fully jacketed ogival nosed bullet with only a base recess.[318]

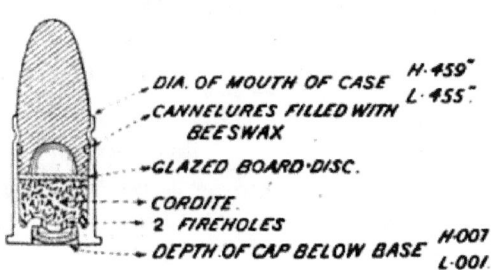

45. Webley Cordite Mark III cross-section. G.B. 1898: 14,659

Ammunition collectors recognize that for years afterward the Webley manstopper bullet was being loaded into cartridges by ammunition suppliers.[319]

Webley patented the manstopper in the United States but only with one base and recess shape in the illustration.[320] The Ideal Manufacturing Company of New Haven, Connecticut evoked the manstopper label in their advertisements for a .38 caliber bullet with a nose recess but a flat base.[321] Ideal contended that the tight fit of the base in the gun's barrel was sufficient for a gas check. But Ideal included designs like the Webley original in their large catalogue of bullet molds for individual casting.

The Webley design became part of the large repertory of bullets privately produced by hunters and battle reenactors using muzzle loading guns.

A 1983 patent has both nose and base recesses and is encircled by gas check and lubricating bands.[322] The patent does refer to the U.S. Webley patent but makes no reference to bullet expansion, which apparently was not an element in reenactments. A contemporary manual for those who cast their own bullets, gives instructions on casting the Webley manstopper.[323] The author advises that they are close range bullets to be fired with light powder. The nose cavity tends to fill up with cloth and other materials. In the recreation environment of bullet swaging the social and political essence of the expanding bullet is gone, and its former urgency as a check to fanatics ceases to be apparent to the maker and user.

In the same bullet patenting free-for-all that yielded the Webley manstopper, Leslie Bown Taylor contributed bullet designs that skirted the other patents. Taylor's Patent No. 14,659 entered on July 4, 1898

---

[318] Great Britain.War Office. Woolwich Ordnance College (1902: 407)
[319] Labbett (1993: 213)
[320] Patent No. 634,383. October 3, 1899, by T.W. Webley.
[321] A New Bullet, *The Pacific Coast Magazine* 5(1904): 513
[322] Patent No. 4,417,521. October 29, 1983, by Ronald R. Dahlitz and assigned to the Buffalo Bullet Company.
[323] Corbin (2012: Chapter 3)

presented forms almost architectural in their construction of metal shapes made to expand regularly at high speed impact.

46. Leslie Bown Taylor bullet patent diagram. G.B. 1898: 14, 659

The noses aren't recessed but are shaped to fold out and obturate the barrel. Too wide a base spread risked leaving lead in the barrel and causing the front portion of the bullet to separate with considerable loss of kinetic energy. Taylor's solution to this problem was to bond the bullet to the casing with an encircling crimp (Patent No. 13,466 from 1899).

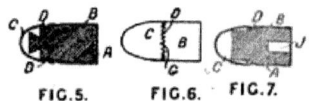

47. Taylor bullet patent diagram. G.B. 13, 466, 1899

Taylor patented the same "compound bullet," formed of two distinct metals bonded together, but with an ogival nose only in the United States in 1902 (Patent No. 693,534).

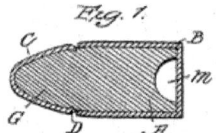

48. Taylor compound bullet patent diagram. U.S. 693, 534, 1902

At the time of the patent Taylor was an officer, one contemporary called him the "ruler" of the Westley Richards Company, a Birmingham manufacturer in business since 1813. The company was known for its hand tooled (rather than machine stamped) firearms, especially rifles in a size range from .401 caliber to "miniature." Several of Taylor's patents were for improvements of rifle loading and firing mechanisms.

Westley Richards operated shooting ranges where buyers, members of the nobility, businessmen and political leaders and foreign dignitaries, tested the power, accuracy and handling of the individually crafted pieces. The company served the occupation of shooting, a manly accomplishment that might extend to wartime shooting but was mostly focused on hunting and competitive marksmanship. Taylor was quite ready to enter into discussions on guns and ammunition in journals for sportsmen as they served to promote company products. In 1913, he co-authored a history of the company.

The Compound Bullet was a hunting bullet meant to achieve the most efficient kill of game animals without littering the carcass with fragments. It was ancestral to one of the most successful hunting bullet designs, trademarked Core-Lokt by the Winchester Company, by the 1960's when the bullet went into production a subsidiary of DuPont. Available in pointed and ogival soft point and hollow point versions, the Core-Lokt was sold for its accuracy. It mushroomed to double its original caliber and achieved good penetration, all attributes of a bullet with high stopping power. Except that it was and is a hunting bullet not for military use.

If production of the Webley manstopper was stopped and not resumed and the best dumdum was restricted to hunting game, how was the dumdum niche filled for bullets aimed at humans? What bullet designs satisfied the need for bullets that lacerate the target to a sure stop?

A principle proposed during the early dumdum age to make bullet expansion regular was the bending or slitting laterally of the jacket at the nose. When the prepared bullet meets the resistance of an outer layer, the divided sections fan out regularly accompanied by the front of the lead core. The spin imparted by the barrel rifling makes the leaves act as cutting blades as they enter the flesh. These illustrations from an 1899 patent (No. 4426 by H.W. Gabbett-Fairfax) give superior, lateral and cross-section views of two variant designs.

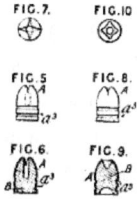

49. Gabbett-Fairfax bullet patent diagram G.B. 1899: 4426

The aspiration to uniform expansion was not realized in practice. By the 1940's slit nose bullets were being grouped with soft nose and hollow nose types as mushrooming bullets. A diagram in a wartime newspaper article even treats making slits in the jacket as the chief method of making service bullets into dumdums.[324]

---

[324] *PMDaily* (New York, New York), July 29, 1942: 6.

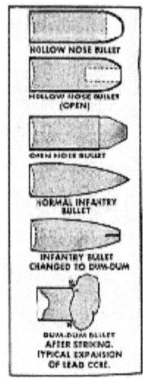

50. Making service bullets into dumdums. Fn 321

All of these types were incorporated into models sufficiently modified and requiring enough new materials and features to receive patents. Patent law and production systems encourage slight reworkings of existing models while reaching toward breakthroughs.

The evolution of expanding bullet design was guided by the desire for a way to securely engineer bullet stopping power that had motivated the invention of the dumdum. Containing this desire were the legal and humanitarian considerations that constrained bullet production policy and deployment from the late 19th century onward. Any attempt to assess what a design actually could do to a target and how it was any different from other bullets was enmeshed in the conflict between urges to inflict decisive damage to a determined adversary and the wish at least to appear humane and law abiding.

The rhetoric that James W. Huffman used to state the utility of his 1998 "Obstacle Piercing Frangible Bullet" (U.S. Patent No. 5,763,819) is that of the "charging fanatic"-stopping dumdum. Driven by adrenalin or drugs, the attacker might survive several conventional bullets and would only succumb to a projectile capable of "deep incapacitating penetration." This being the purpose of Huffman's own bullet, he reviews others recently or currently made with this in mind. Though the descriptive name of his own bullet is not meant to suggest use against humans but rather enhanced penetrative capability in general, the bullets he reviews (and his own) are for law enforcement.

One of these bullets represents a further step in the evolution of the expanding bullet totality: the Winchester Black Talon SXT-TM (Supreme Expansion Talon). The bullet's name, trademarked by its producer, the Olin Corporation, and its logo, curved black talons grasping the name block, metaphorically reflect its features.

A reverse tapered hollow point crimped vertically along 3 diameters, the copper jacket was structured to peel back to form a rotating 6-bladed petal on impact with the fluid mass of the target. Tested as a bullet that could pass through safety glass without fragmenting-in other words as a pistol bullet for police use-the Black Talon stretched out its petals when fired into ballistic gelatin, the closest approximation of flesh density.

The talons did not seize but sliced through the intervening flesh churning an expanded cavity through muscle, vessels and bone. The bullet nose was coated with a patented black lubricant that sped them through the gun barrel. Ricochets and fragmentation, always of concern to police policy makers, were minimized by the coating. The Black Talon series were handgun bullets available in several different calibers.

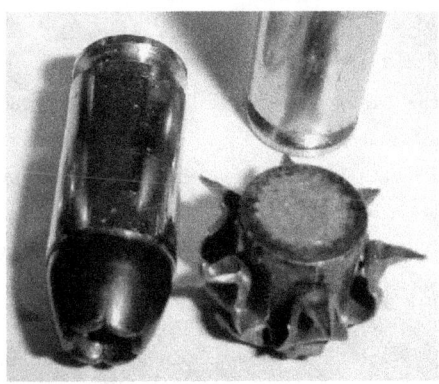

51. Black Talon bullet in cartridge and expanded form.

The performance of the Black Talon bullet depended upon an engineered coordination of bullet design with the expenditure of kinetic energy in the fluid medium inside the skin. The manufacturer's publicity encouraged potential users to expect consistent results, the dumdum proposal further formalized. The gunpowder charge and bullet weight were balanced to prevent perforation. The pointed petals of the copper jacket were reinforced to retract uniformly without deforming in response to fluid resistance after their initial expansion. This reduced resistance to the bullet's further penetrated. The shape of the expanded bullet might encourage turning and tumbling as the energy diminished.

Some presentation of the Black Talon in specialty magazines praised it as a hunting bullet.[325] It was adopted by police departments nationwide as a service bullet. The Black Talon imagery, the pictures of the (ballistic gelatin) bullet pointed talons extended, soon became a public liability.

On July 1, 1993 Gian Luigi Ferri walked through the offices of the San Francisco law firm Pettit and Merrill firing bullets from handguns he carried. Ferri shot anyone he happened to pass, killing a total of 9 people (including himself at the end) and wounding 6. Ferri's connection with the firm was far in the past and he did not seek out individuals. No one on a hit list he carried with him worked at the 101 California St. location, which then lent its name to the shootings.

The guns and magazines in Ferri's possession, and the ammunition he fired, a combination of standard and Black Talon bullets, became the subject of litigation and legislative initiatives.[326] The initial newspaper

---

[325] Petzal (1993)
[326] Kirk (n.d.)

reports on the shootings did not name the guns or bullets but somewhat later turned to the bullets.[327] As the vicious-looking expanded form of the Black Talon was described and depicted, the public outcry moved legislators to propose an outright ban on the bullets or a heavy tax on their purchase, one opportunity in the broader gun control movement. Remington-Olin undercut this threat to their business by withdrawing the bullets from sale to the public.[328] Only police departments could buy them.

A shooting on Long Island Railroad cars on December 28, 1993, one month after Olin withdrew the bullets from general sale, was a reminder that they were still in supply

Emergency room physicians and surgeons expressed apprehension at probing wounds for Black Talon bullets. Their gloves and fingers might be sliced by the sharp petals before the bullet came into sight, exposing them to HIV and other blood contaminants.[329] Procedures for the forensic recovery of bullets from bodies and crime scenes had to be geared to the hazards the Black Talon posed.[330]

Yet in 1994 Boyd Stevens, the San Francisco medical examiner at the time of the 101 California St. shootings, presented a paper on the wound ballistics of the Black Talon based the examination of those killed by Ferri.[331] He found the wound patterns "unremarkable" and concluded that the Black Talon bullet wounds were not readily distinguishable from the wounds left by the conventional rounds they accompanied.

During its brief period of general availability by name the Black Talon evolved in a manner similar to the dumdums and manstopper. It was conceived as a means of stopping a raging antagonist. It was denounced as an extraordinarily cruel means of dispatching someone. A police ballistics officer interviewed at the time of the bullet's withdrawal observed that the Black Talon imagery was a factor in the public reaction and calls for a ban.[332]It was renamed Ranger SXT and continued to be sold and used both by the general public and the police.[333] It also served as the basis for PDX-1 that the FBI adopted. Like its predecessor expanding rounds the Black Talon design persisted whatever it was called. Other very similar bullets appeared on the market.

Expanding bullets are sometimes described as exploding bullets but exploding bullets are a distinct category of projectile. Bullets constructed to explode, not just expand, once inside the living targets were, as they had been since their first use in the mid-19th century, hollow-point bullets with an explosive charge inserted. They are distinct from frangible bullets, which are made to disintegrate in the body but not as the result of an explosion. Attempts to prove that the enemy was making and distributing exploding bullets were a subroutine of World War I dumdum accusations.

Handgun and rifle bullets made to explode inside a target are hollow points with a minute container of explosive inserted in the recess which is then plugged with a nonmetallic material. The kinetic energy of impact is expected to force the plug into the explosive and cause it to detonate. International agreements since the 1868 St. Petersburg Declaration forbade the anti-personnel use of exploding bullets. They continued to be made and used for stated and unstated purposes.

[327] Victims of Caprice in Deadly Rampage, *The New York Times*, July 7, 1993.

[328] Smothers (1993)

[329] High Tech Death from Winchester, *The New York Times,* November 11, 1993

[330] Russell, et al. (1995)

[331] Stevens (1994)

[332] Supenski in Smothers (1993)

[333] Firearms Tactical Institute (1998) Winchester Ranger Talon (Ranger SXT/Black Talon) Wound Ballistics

In 1979 Bingham, Ltd., a firearms and ammunition maker in Norcross, Georgia (USA) began the sale of an exploding bullet for handguns they named the Devastator. The name was not trademarked for this bullet: there already was a line of hollow point bullets by that name, the Lyman Devastator, 429640 in the Ideal/Lyman numbering system. The Devastators Bingham sold were hollow point lead bullets containing a lead azide capsule sealed with lacquer and loaded into a .22 cartridge. The company's publicity stated that they were for small game hunting and target practice. The Velet Cartridge Company of England sold a similarly constructed bullet they called the Exploder.

The bullet might have remained among the many types of projectile used by hunters unremarked by the general public except that they were fired at President Ronald Reagan by John W. Hinckley, Jr. on March 30, 1981.They then became a stage in expanding bullet evolution.

Reagan's press secretary, a District of Columbia police officer and a Secret Service agent were struck by bullets before Reagan himself was hit by a ricochet as he was hustled into his limousine. Rapid transport to a hospital and surgical removal of the bullets probably saved lives. The surgeons did not know that the bullets were explosive until FBI investigators found an empty box labeled "Devastators" in Hinckley's hotel room.[334] They then concluded that the bullet that entered the skull of James Brady, the press secretary, had exploded, but none of the others had. The concern that unexploded bullets lodged in a body might detonate during diagnostic imaging or surgical procedures was expressed at the time and repeated in subsequent articles in medical journals.[335]

In October, 1981 the federal Bureau of Alcohol, Tobacco and Firearms sent Bingham, Ltd.
a letter warning them that they could not continue producing the Devastator bullets without an explosives license. Obtaining such a license would increase production costs and therefore the price. Facing diminishing sales Bingham sought a court motion recognizing the explosive insert as a component, exempting them from the license requirement. The Eleventh Circuit Court of Appeals in February, 1984 sustained the district court's denial of the order.[336] The diminished supply of Devastators was abetted by imports.

The Reagan assassination attempt was prominent evidence that often the bullets did not explode, and did not cause inevitably lethal damage when they did. The feared instances of emergency room catastrophe when unexploded bullets burst during procedures did not materialize. A sequence of letters in the *Journal of Clinical Pathology* in 2004 culminated in a cautious dismissal of medical staff unease at the possibility of encountering an unexploded bullet in a gunshot wound.[337]

Exploding bullet fears arise from the same platform of danger arising from malice formally built into a bullet. As a shooter can administer additional pain to the victim by making a bullet that expands and tears and explodes, he can extend that pain to rescuers and medical staff. The danger of the bullet is equated with the intent of the shooter as imagined by those who might be affected and not by the documented effects of the bullet itself. A bullet that is made to blow apart a small animal must expand beyond that in a human.

---

[334] Taubman (1981)

[335] Knight (1982)

[336] Bingham, Ltd v. United States, 724 F.2d 924

[337] Swift and Rutty (2004)

The exploding bullet has evolved along with the other expanding bullets into an intention brand when aimed at humans. As with the dumdums and their predecessors they are shot across a divide with the force of imagination, but their real effects are a matter of conditions. Breivik accomplished his murders with explosives and projectiles that had their lethal effect because of the circumstances of their detonation and targeting, not because they physically embodied his antagonisms and desperation. The malign intent does not tear the flesh apart. The bullet only pierces it.

15. Exhibition of the Moment

The 1872 album of 12 photographs entitled *Tiger Shooting* illustrates each phase of what was then considered sport, from setting up camp and setting out the bait carcass to the skinning of the tiger.[338] All of the photographs taken by W.W. Hooper and V.S.G. Western for this souvenir series were staged. An artfully positioned dead tiger was the object of the shooting by gun and camera. Most of the tigers in photographs by European and Indian photographers during this period were dead specimens on loan from hunters and rulers. Living tigers appeared only in sketches and paintings.

Lt. W.W. Hooper was ambitious to capture the moment of death even this early in his career as a photographer. A staged series was the best he could do with tigers. To picture the facial expression at the exact moment the bullet entered the human body, Hooper, then a colonel and Provost-Marshal of Burma, delayed the execution by firing squad of three dacoits to allow him to snap the shutter when the guns were discharged. He did get the photograph, and nearly was court-martialed when the viceroy learned of the circumstances. The existing prints show the dacoits in the distance.

Grattan Geary, the editor of the *Bombay Gazette*, dismisses any inhumanity in Hooper's "desire to photograph the Burmese when struck by bullets." [339] It was what "almost may be regarded as a passion for securing an indelible record of human expression at the supreme moment." In confirmation of this passion of Hooper's, Geary recalls an incident.

> It is related of him that when a sepoy went shooting at large at his officers
> and comrades that he ran out with a photographic apparatus and aimed it
> at the sepoy, who was taking aim at him. The homicidal soldier was struck
> at the instant by a bullet from another sepoy, and Colonel Hooper obtained
> his negative.

This casts light on Hooper's other reported acts of daring: going out onto the battlefield carrying a camera One purpose of this must have been to catch the faces of men being shot.

Hooper's main photographic work was collections of portraits of tribal people, including arranged group scenes of victims of the 1876-78 Madras famine. These people may also have been about to die and they certainly were suffering though without the drama of receiving a bullet. Hooper's sudden moment of death shots if they remain in any collection have not been catalogued or published. The renegade sepoy and the shot soldiers are mental images known only from the account of their making. The gallery of faces and occupations in Hooper's collections can be viewed in published prints, but not the sudden death shots.[340]

The moment of death photographs are the imagistic kin of the shape changing bullet. The bullets are fashioned to take on a decisive shape on entering the body, an instant reflected in the suddenly changed physiognomy of the recipient. The only account of the bullet entering the body was in the anatomy of the wound and the spent bullet recovered. Hooper's quest to photograph the face at the moment was the only

---

[338] Hooper and Western (1887)

[339] Geary (1887: 242-43)

[340] On December 3, 2012 *The New York Post* printed images taken by a freelance photographer of a man who had been pushed off the station platform onto the subway track as the train approached. Gerda Taro's 1936 photograph The Falling Soldier was determined to be of a soldier losing his footing, not after being shot.

record of a body change that conceivably could be obtained at the time. This was a portrait of a man at the moment of bullet-caused injury and possibly death. The shape changing bullet design was an attempt to package a physiognomic force in the bullet.

Making a bullet that would by its shape lead to a predictable ballistics from firing to wound was the aim of bullet design. Visualizing the bullet's course into and within the body itself was a major factor in understanding how bullet shape could change and could be made to change. Following the bullet into the wound led away from the external effects of receiving the bullet, from facial expression and body position that communicated the sensations of the recipient. While attempting to capture these nervous or humoral results of the bullet's arrival, photographers like Hooper initiated a neutral record of that event. The emotionally difficult face was about to be dissolved into the medically useful interior of the body. Improving bullet design by following the bullet into and through the flesh also took attention away from the shock and pain in the face responding to the more lethal design entering the body.

Ballistics studies of bullet force and bullet path entailed firing bullets into materials of consistent texture and density. Graphics like this one from a late 19th century newspaper article on rifles dramatized the increase in kinetic energy attained by the composite effect of gun and bullet design.

52. Bullets of different kinetic energy fired into the same materials. Fn338

Comparable experiments with flesh relied on living and dead animals and human corpses. Current equipment did not allow photographic recording of the process or the wound track. The mental image of a bullet's performance in a target was based on assumptions about changes observed in the recovered bullet and not on visual verification of bullet passage. Only the gross results of warfare and the occasional opportunity for an autopsy provided information and verbal descriptions of the bullet's course. The science of photography had not progressed beyond attempts to record physiognomy at the moment of bullet strike.

In 1897, for instance, the Ordnance Department of the U.S. Army undertook tests comparing a manufactured .30 caliber soft nose bullet with the service bullet.[341] Controlling for range, the bullets were fired into wood, horse flesh and other materials and the accuracy of the shot, the severity of the wound and the degree of mushrooming were recorded. The advantages of the soft nose bullet did not offset its disadvantages and the service bullet was judged superior.

Four years later a Navy captain stationed in Cuba tested unidentified soft nose bullets against service bullets by firing them into a freshly killed horse "while the flesh was warm and practically living."[342]

> Shots were fired so as to pass through the full thickness of both hams from side to side to side, piercing the skin four times; the bullets were not recovered. The channels of the two bullets were then immediately laid open side by side, and no difference in the track of the jacketed and soft-nosed bullet could be noted; nor was there any difference in the size of the final wounds of exit. It is presumed that the photographs showed the result of experiment on cold, stiffened flesh, and that resistance offered in that case was sufficiently great to set up the bullet.

Foltz hoped to obtain results reflecting the passage of bullets through the living flesh but was only able to document photographically the perforating wound track in cold, more resistant flesh. Yet under those conditions the track of the soft nose bullet which must have "set up" was no wider than the track of the service bullet. Foltz concurred with the findings of his Army counterparts and found that the soft nose bullet was no more destructive than the service bullet. The static documentation of the exposed wound tracks through dead flesh was the best evidence he could obtain.

The method of locating a bullet in a wound was probing, with fingers or an instrument, the same as it had been in Ambroise Paré's time. Opening a wound track surgically just to search for a bullet risked causing further injury and blood loss without removing the bullet. Bullets not quickly located were left in the body, and many veterans carried more than the memories of their battles. X-rays found use soon after the announcement of their discovery in January, 1897, in determining the position of invisible bullets in the body. By providing shadowgraphs of the bullet mass and associated damage to bones and possibly organs, X-ray imaging guided medical treatment of wounds. It also gave the picture of the bullet track closest to the real time of the wound.

The first X-rays of soldiers' bullet wounds were taken in Italy a month after the wounds were received during the Abyssinian campaign. The bullets could have been left in the body; removing them safely was a new option. Yet field surgeons saw the advantage of having X-ray apparatus in battlefield treatment tents, to find bullets that could not be left. Surgeon-Major Walter Beevor of the Royal Scots Guards purchased an X-ray apparatus and saw to its transport to the field of action in the 1897-98 Tirah campaign. The British India army troops fired dumdum bullets at the Afridi foe, and received in return a variety of balls and slugs.

---

[341] U.S. Army. Ordnance Department (1897: 105-09)
[342] Foltz (1901)

A "Ghoorka" (Gurka) on the British side was shot in the back of the thigh.[343] The surgeons tried probing the wound without locating any bullet yet the loss of motion suggested that there was a foreign body present. They could only wait until the inflammation and swelling subsided, which might not have happened soon enough to save the man's leg. A field X-ray showed the bullet's shape and position and allowed its course to be surmised. It had entered diagonally from above and traveled downward and inward, struck the bone deforming as it rebounded in a channel of its own.

53. X-ray of Gurka wounded in back of thigh. Fyfe (1900: 839)

These X-rays were the first military field X-rays printed in a magazine with a general readership. The article on military X-rays in *The Strand* was preceded by a sequence of photographs of a jugglers' dinner party with plates, cutlery and bottles suspended in the air. In the field X-rays internal body structure, mostly that of bones visible, was uncovered under assault by the enemy bullets caught where they had lodged. The article begins with an account of General Wodehouse in the surgical tent withstanding a sudden assault by Afridis, a model of heroic manhood with the X-ray of the bullet-puncture in his leg as evidence. In some of Beevor's published pictures the whole bullet is halted in its approach to a bone. The wound tracks are not visible as they became when opened surgically guided by these photographs, but the destructive effects and shape change of the bullets tells their story in the body.

The fortitude of the fighting man who endured punctures and lodged bullets now seen in photographs was not the same as the heedlessness of ghazi charging a barricade. The publication of battlefield X-rays, taken in the midst of the fighting, confirmed the endurance of our own soldiers in the face of a relentless enemy. Whether or not the track was visible, the career of the bullet was tangible in photographic representation. The same sense of raider-halting destructiveness that underwrote the dumdum was contained in the bullet picture.

The X-ray apparatus gave pictures of bullets entering friendly soldiers as the preparation of dumdums gave pictures of bullets entering the enemy. The apparatus was conveyed into the field for the British punitive expedition to the Sudan. Military medical services published accounts of ingenious portable set-ups, ways of generating the needed current and viewing the fluoroscope screen in brilliant sunlight. They also

---

[343] Fyfe (1900: 539)

published an increasing diversity of bullet wound images first in special reports and scientific journals, then to accompany magazine and newspaper articles.

The American invasion of Cuba gave medical authorities the occasion to question the forward field stationing of X-ray equipment advocated by Beevor.[344] Captain W.C. Borden, the surgeon who prepared the general report on the use of the Roentgen ray during the war, argued that stationing X-ray equipment at the front lines was unnecessary given the rare need to remove lodged bullets immediately, and dangerous because its presence encouraged surgeons to operate under conditions that might not be aseptic.

Borden tabulated statistics from wars of the previous century, before and after the discovery of X-rays, to underline his contention that it lessened mortality to use X-ray apparatus and to locate it in base hospitals and on hospital ships. The percentage of the wounded who died after coming under treatment for the American civil war was 10.96; for the Tirah expedition it was 10.44. But the death rate for the French during the Crimean war (1854-56, no X-rays) was 10.90 and for the Prussians during the Austria-Prussian war (1866, also no X-rays) war was 10.59. The Spanish-American war rate for the Americans, with base stationary X-ray equipment was 6.44. The low rate the British achieved with field X-rays at Tirah was not much lower than had been achieved without X-rays in other wars. The Americans believed they had proven that X-ray equipment at stationary hospitals was the optimum arrangement for a significant reduction of the mortality rate. Other factors may have been responsible for the differences in mortality, but at the time the trend seemed to be in favor of stationary hospitals.

The report is an atlas of radiographic plates of the gunshot wounds to different parts of the body in many cases before removal of the bullet. The wound track is not visible but was inferred by the relation of the entrance wound to the location of the bullet, and its shape in relation to visible bone damage. The Mauser and other bullets deformed by ricochet or passage into and through bone. Techniques had been developed to determine the three-dimensional location of the bullet in the limb from its position on the two-dimensional radiographic image.

For instance Plate V (Case 6) revealed the stopping place of an undeformed Mauser bullet at a right angle to the femur.

---

[344] U.S. Department of the Army. Office of the Surgeon General (1900: 18-19)

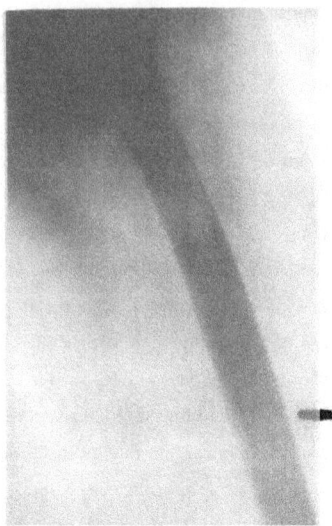

54. Radiograph of an undeformed Mauser bullet perpendicular to the femur. Fn342: Plate

Drawing a line on an anatomical figure between the entrance wound and the bullet's radiographic position outlined the wound track and showed that the bullet had ricocheted and tumbled internally to its changed direction without visible deformation.

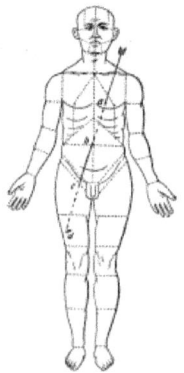

55. Bullet trajectory through model. Fn342

The radiograph disclosed what would have remained hidden from the conventional probe. Disturbance of the wound track by probing threatened to cause further damage without locating the bullet.

In some of the cases a bullet perforated a limb and exited leaving no radiographically visible damage. Then the X-ray obviated the need for probing. In other cases the only pictorial evidence of the bullet's passage was a bone fracture. The Mauser bullet in Plate XXVI has caused a comminuted fracture of the right femur and halted having deformed to a mushroom shape without exiting.

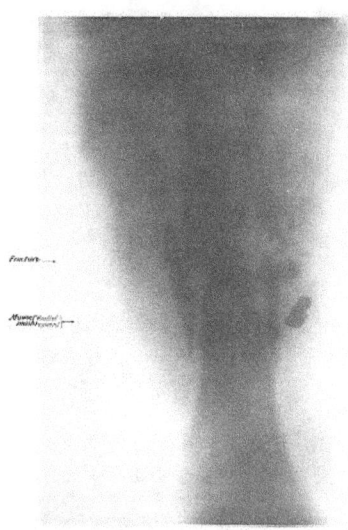

56. Mauser bullet deformed after fracturing right femur. Fn342: Plate XXVI

The radiograph projects the spatial and temporal disposition of the bullet. It gives physical evidence of the relationship between final bullet shape and the force of its projection along the line it followed through the man's thigh. The bullet is distinguished from the broken bone and the swollen tissue as a foreign projectile that may have carried infectious foreign matter into the body.

The expanded state of the bullet was a consequence of its contact with a long bone while still driven by enough kinetic energy to cause the fracture and be spread by the resistance. Captain Borden includes a plate of 6 Mauser bullets exhibiting altered shapes caused by ricochet against stone or metal outside or by contact with bone inside. He does not include bullets that were designed to deform and augment the wound, because no such bullets were encountered. X-rays depict the moment of the wound in terms of bullet shape and course as having numerous possibilities, one of which is deformation.

In one of the his medical reports on the British war in South Africa (Boer War) two years later the surgeon Clinton Dent cautions that comminuted bullet fractures are not invariably the result of bullets prepared in violation of the Geneva Convention [Hague Convention].[345]

> That bullets are in some instances "doctored," and that forbidden kinds of bullets are occasionally used, is beyond all question true, for such bullets have been removed; but it is but just to state that instances of the kind are of extreme rarity. In any kind of contest the mere suggestion of unfair play is apt to make people lose their heads quickly, and when in warfare the question of infringement of methods sanctioned by civilized communities is raised, exaggeration is prone to run riot on very slight evidence, and on very insufficient data.

---

[345] Dent (1900a: 969)

Dent then refers to photographs of unfired and recovered Mauser bullets that have deformed as the result of impact against the bone. That possibility, also illustrated by radiographs of the deformed bullets prior to removal, can make it appear that the Mauser bullets had been altered to expand. Dent compares cross-sections of the thinly jacketed Mauser bullet (A) to the thicker-jacketed Lee-Metford (B).

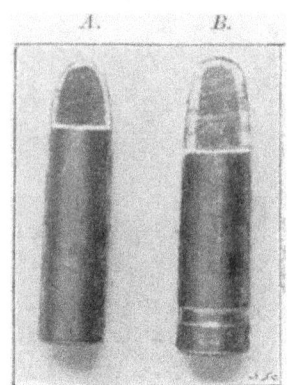

58. Nose cross-section of Mauser bullet (A) and Lee-Metford bullet (B). Dent (1900a: 969)

The Mauser's jacket makes it more likely than the Lee-Metford to deform when striking a solid surface: another explanation for the mushroomed bullets sometimes found. The speed of the bullets due to the powders used causes them to pass through their targets rather than expending their energy within the body and coming to a halt bearing infectious matter. "A Humane War," Dent terms the conflict in another article,[346] the antithesis of a war waged against uncivilized savages with dumdum bullets. The evidence of the wound tracks and recovered bullets marks this as a conflict in which the combatants are not out to cause great pain, only to remove the opponent from the field.

The Spanish-American War had already given the lie to assertions that the lower caliber, high velocity, metal jacketed bullet was "more humane" than its predecessors.[347] The exploding bullet label was applied to these bullets from the damage they did. Exempting bullets not explicitly designed to expand or explode from savage intent causing more severe wounds was part of a developing account of the bullet in the flesh.

Following earlier researches, Charles Woodruff, an army medical officer stationed in New Orleans, in 1898 published an article accounting for the explosive effect of the high velocity bullets. It is analogous to the cavitation rapidly moving bodies induce in incompressible fluids.[348] A bullet entering flesh opens a temporary cavity filled with hot gases that induce necrosis and set the stage for infection in the surrounding tissue. The permanent cavity (the bullet track) would seem to be the collapsed result of a forward explosion. Bullets not constructed to expand or explode did damage from the forceful release of hot gases, as in an explosion.

---

[346] Dent (1900b)
[347] Cirillo (2004: 49)
[348] Woodruff (1898)

The discovery that low caliber, high velocity bullets acted in phases called for ballistic imagery that isolated those phases. Shooting a bullet into a solid material would not in cross-section distinguish critically between expanding bullets of different types and bullets impelled by the same powder but not designed to expand. The question whether there was a distinction between the tracks of expanding and non-expanding bullets could not be answered by existing terminal ballistics imaging. To exhibit the moment of the bullet's passage required sequencing it.

The rapid introduction of X-ray equipment into military medicine at the end of the 19th century did not precede a wholesale adoption of the technology and associated diagnostics in the following years. Louis La Garde, who as an army medical officer had supervised a unit using the equipment to locate bullets in the Cuban theatre of the Spanish-American war, included in his treatise on gunshot wounds a recent (circa 1916) X-ray showing a bullet lodged (1864) in the shoulder of a Civil War veteran.[349] During the fifty or so years the man carried the bullet in his shoulder he experienced some pain but was not prevented from performing his usual work. La Garde compiled X-rays from several wars for the first edition of his book. The method only came into its own with the outbreak of the World War I in 1914, when the Civil War bullet was imaged still in place in the living man.

A sufficient fund of X-rays and photographs had accumulated for La Garde to juxtapose frames of different types of bullet fired into the same part of the body from the same distance. In order to augment his comparative repertory he had conducted experimental firings into suspended cadavers. His book follows earlier work in forming an atlas of gunshot wounds by body region. At the beginning he recognizes the current theory of wound ballistics and the explosive effect of high velocity bullets in flesh, and he attempts to observe that progress in each of the regional instances.

The closest he can come to exhibiting the time series of a bullet is by illustrating single instances in which the entire career of a bullet of specified composition can be documented and analyzed. This photograph of the right thigh of a suspended cadaver shot with a pointed bullet at the "simulated range" of 100 yards.

---

[349] La Garde (1916: 39)

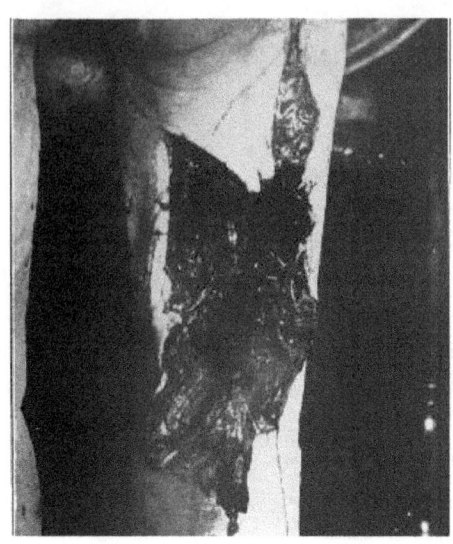

58. Thigh of suspended cadaver shot with pointed bullet at simulated 100 yards. La Garde (1916: 59)

The entrance wound, La Garde wrote, was the same size as the bullet; the exit wound shown in the photograph was large enough to admit a fist. The bullet had fragmented the middle third of the femur and after its explosive exit turned and hit a nearby barrel side on. The bullet was neither constructed to expand nor to explode yet this observed breakdown of its phases demonstrates both. It entered the fluid medium of the thigh, expended its kinetic energy in fragmenting the bone and forming and explosive temporary cavity made permanent on the surface of the exit wound before tumbling into a solid outside surface.

A Colt automatic pistol was used to fire a .38 caliber full-jacketed bullet into a tibia from 37 ½ yards (left) and from close range (right).

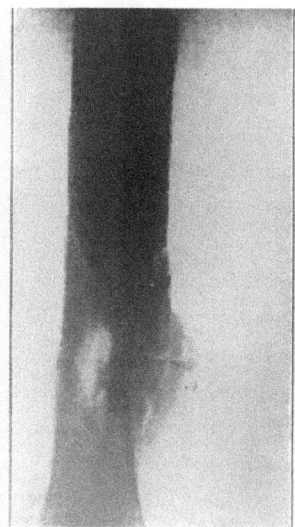

The same pistol fired a soft nose .38 caliber bullet into a tibia from 5 yards.

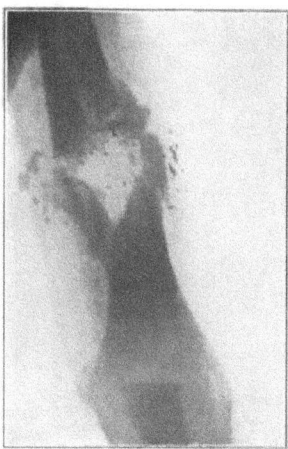

The fully jacketed bullet caused greater damage to the bone at a distance than at close range; a soft nose bullet of the same caliber caused the most damage at close range. La Garde's X-rays illustrate what was long suspected: the kinetic energy of the fully jacketed bullet carries it through bone when fired from nearby. The soft nose bullet shatters the bone shaft, leaving a large gap with extensive fragmentation. Its open lead nose has mushroomed on meeting the fluid, slowing it down and magnifying the extent of the perforation. It has rendered the limb unusable. The comparative severity of the wound is a visual index of the effects of presence or absence of the jacket upon the phases of the bullet's action. La Garde exhibits the bullet's moment phases without capturing the bullet itself in motion. The bullet parameters affecting wounding capability and stopping power are displayed at the second of wounding. The record of the bullet's moment no longer includes or refers to the facial expression of the recipient, or to his feelings.

La Garde's purpose was not to study bullets but to anticipate the medical services needed, the need for battlefield evacuation and the frequency of disability, when certain types of bullet are in use. The number of examples has permitted the construction of a comparative visual database that amounts to an effective analysis of the bullet's terminal ballistics in the human body.

This level of documentation did not exist for most parts of the body: bone and organ injuries were the most graphically visualized by X-rays. The comparative X-ray technique that yielded this imagery was by the time of La Garde's second edition not the most iconic for the traveling bullet. In 1909 a German physicist, Carl Julius Kranz, demonstrated a ballistic kinematograph, a film camera coordinated with a rapid spark flash that broke down high speed events into a sequence of distinct cinematic frames.

The first photograph of a bullet in flight had been taken in 1880 with a spark flash. The ballistic kinematograph recorded the material events of a bullet penetrating materials such as bone and water-filled bladders analogous to organs, as in these frames of an infantry bullet disintegrating bone.[350]

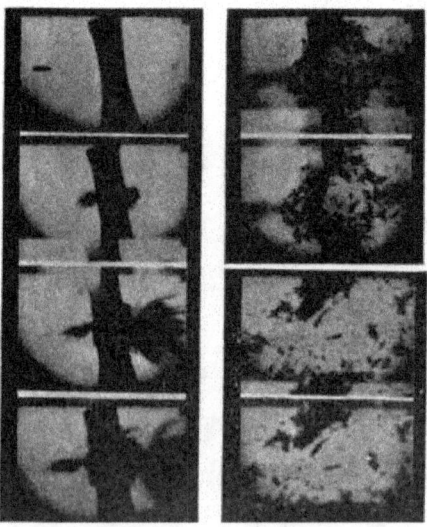

59. Ballistic kinematograph of an infantry bullet shattering bone. Fn348

As popular accounts pointed out, these brief films contributed to the performance analysis of firearms and bullet components. For instance they allowed technicians to assess the performance of gunpowder by tallying

---

[350] Lehmann (1911: 112)

the proportion of unignited grains emerging from the gun.[351] Military surgeons did not find them as useful as X-rays in examining the course of bullets through the body since kinematographs could only be made of events staged in the open. The differentiation of bullet types achieved by comparing X-rays of injured limbs could not be matched by the ballistic kinematograph. Bullets could be compared cinematically but not in their changes and effects on flesh living or dead.

Improvements in X-ray and kinematograph technology and the introduction of new technologies for the study of bullets in materials have not yielded imaging of the bullet's progress through flesh. Translucent ballistic gelatin allows tracing the path of a projectile through a uniform medium corresponding in density to the body interior. Stroboscopic and stop motion photography in their digital forms gathers microscopic details of the phases of travel. The imagery of the bullet's stages of travel relates and its condition at each of those stages equates bullet preparation with result.

Visually following the bullet in the body has not been attained in reality, but it has been attained in fantasy. Staging the bullet passing through bone for the camera has merged with ever more sophisticated cinematic and videographic techniques to simulate what has never actually been observed. Computer graphics interface allows us to follow the projectile into the skin through blood vessels, organs and bone. These animated bullets always are generic; soft nose and manstoppers are not distinguished from fully jacketed service bullets.

Models made of gelatin composing artificial organs receive sword thrusts and bullet wounds with equanimity, to help evaluate the relative effectiveness of weapons. The gelatin can only reflect the fluid dynamics of the thrust and not the wounding potential. Facial expressions at the time of bullet entry into flesh have been mimicked by actors for film and video representation, and the passage of the bullet is followed in computer graphics for forensic reconstructions.

Hooper's quest to seize the facial expression and muscle reaction at the moment the bullet enters the body has been lavishly fulfilled, but only as fiction. The picture of bullet performance that had been in mind since the first bullets shaped to a purpose has been realized without the participation of the person shot. He or she is present only as body systems being curtailed by the entrance and passage of the bullet. The added potential for pain in the structure of the bullet is secondary to the bullet's course in the body all of which, following from the first dumdums onward, is referred back to societal and then to personal defense.

---

[351] Long (1909)

16. Personal Defense

In a current (2012) hunting outfitter's catalogue a line of handgun bullets is advertised as "the pinnacle in self-defense ammunition." [352] Using the materials incorporated into another "revolutionary" bullet by the same manufacturer (Hornady), the engineers "have achieved 100% reliable expansion when shooting through even the toughest clothing barriers." The solution to the expansion- inhibiting effect of clothing clogging the hollow point of the bullet is a "polymer insert." Meeting FBI protocols for duty ammunition after eight years of research and development has resulted in "intelligent bullets" that adapt to any barrier they encounter. The jacket is securely locked to the bullet core in order to prevent premature separation. The bullet penetrates every barrier from auto glass to plywood and sheet metal. Endowed with maximum velocity and impact by low flash propellants, the bullet exhibits superior tactical performance.

The only mention of "soft tissue" in the product description is to assert that the bullet is designed not to be inhibited by clothing from expanding in soft tissue. The reference to "terminal ballistics" elsewhere in the article maintains the allusion. No other bullets on the two pages devoted to them in the catalogue are aimed at successfully terminating in soft flesh beneath layers of clothing, or after penetrating auto glass and sheet metal.

As other products in the catalogue depend upon the browser having a hunter's sensibility and experience, so the Critical Defense bullet's description sells the product by referring to the anticipated moment of shooting a human attacker. In this moment the bullet penetrates glass, steel paneling and layers of clothing to expand uninhibited in the soft tissue. There is no reference to stopping power. This is the visual moment exhibited by the animated bullet of films and television, not the spent bullet of X-rays, surgical intervention and ballistic gelatin. The selling point that originated with the shaped bullets of Puckle's Defence now offers a functional shape-changing capability to defend a person against an approaching assailant.

These bullets are offered in a range of calibers from .303 to .45, as are others with a similar sales pitch. No one size bullet is preferred over others; any handgun can be made into a personal defense gun. There is a critical industry evaluating which is most suited.

The language of bullet presentation is one expression of an ideology that has assimilated military and police conceptions of shape-changing bullets to the individual bastion promoted by gun marketers. The soldier who aimed at the ghazi headed toward his line and the police officer who aimed at the drug-crazed felon on a rampage have been compressed into a single person with a gun and bullets threatened by motion outside of the self or the residence.

The Federal Premium Personal Defense Handgun Ammunition, a competing line of bullets, has its stopping power due to controlled expansion. The jacket of this hollow point is notched to cause it to expand "hydrostatically" when it penetrates a barrier. It mushrooms until a "hardened central post base stops the expansion." Thus it expends its energy and leaves its weight in the target. "If you hit what you're aiming at, this bullet will do its job."

---

[352] Cabela's Christmas Catalogue-2012: 119.

The makers of this bullet express its purpose even more obliquely than the Critical Defense bullet's makers. No soft tissue is reached; the bullet's stopping power is achieved by the shooter's good aim. The human target of the bullet has become a mass halted by the bullet's umbrella-like transfer of opposing energy. A faceless mass as far as bullet and shooter are concerned.

This vision of a bullet let fly preventively into a malicious other is counterbalanced by the further specification of concealed carry of the firearm. The drive to legalize bringing a hidden weapon into a public place where weapons previously were forbidden is shrewd marketing by the gun industry. It raises the possibility that anyone in a place where guns once were legally and/or consensually impermissible can be a place where anyone can be carrying a gun and therefore you should be carrying one. A fragile civility is enforced where previously it was assumed. The argument that if guns are illegal then only criminals carry them cuts both ways: if they're legal then anyone carrying a concealed weapon is not a criminal by that act alone.

Instruction manuals, in effect firearms self-help books, give advice on the best guns to conceal and the appropriate ammunition, which turns out to be hollow point bullets. One writer assures his readers that the best (legal) defense for a person who defends him or herself with a hollow point bullet from a concealed gun is that the police are armed with these bullets.[353] The equation between personal defense and law enforcement proceeds from the confrontational history of expanding bullets. Concealed carry brings personal defense theme and the hollow point bullet to the individual's private barricade while sanctioning it as a public act.

Hollow point bullets were advertised as small game hunting ammunition from the early twentieth century onward. Whatever their uses in warfare and policing, their consumer identity was tied to hunting. No bullets were attached to self-defense or personal defense in the promotional literature until the 1990's and then, as was seen in the two examples above, their hollow point design was improved by expedients. Personal defense bullets are not associated with hunting bullets.

"Self-defense" is a legal phrase used to counter an accusation of premeditated killing of another person, an accusation of homicide. This phrase appears in newspaper accounts of gunshot deaths from the mid-nineteenth century onward. It was the simplest way to report a shooter's assertion of innocence, of a need to use deadly force against a life-threatening attacker. The attacker's death was not intended, but was the consequence of a spontaneous act to prevent the attacker from carrying out an apparently injurious intent. The self-defense phrase encapsulated the lack of criminal intent that would then become the centerpiece of the shooter's own defense.

A Galveston, Texas attorney named John Lovejoy shot and killed a "Negro drayman" named Aaron Williamson on August 13, 1888. The attorney was descending the stairs from his office accompanied by his partner when Williamson appeared at the foot of the stairs and demanded that Lovejoy pay him for electioneering Williamson had done for Lovejoy's campaign four years earlier. Lovejoy said he would leave the money for him at his office but Williamson insisted on immediate payment. Described as a powerfully built man, Williamson advanced toward Lovejoy easily bypassing the partner who tried to restrain him.

---

[353] Ayoob (2012: Preface)

Lovejoy took out a revolver and fired at the advancing Williamson. "Shooting in Self Defense" read the title of *The New York Times* article on the incident. It did not matter that no gun was found on Williamson's body. Lovejoy was not charged with a crime.

The conditions of the Galveston confrontation paralleled conditions on the frontier of the colonial empires during the same period: the forces of rule facing resisting subjects. Williamson's strength was taken as sufficient cause for Lovejoy to defend himself with deadly force. The stopping power of the bullets did not enter into consideration. Close range and a single attacker were sufficient to end the scene.

When colonial troops firing low caliber bullets faced attacking natives in a mass, special encouragement was required to maintain morale. That took the form of bullet modifications giving technical assurance of stopping power. When the self-defense postures like that of Lovejoy facing Williamson are translated into the attacking native's vernacular, personal defense originates. During the next 100 years self-defense became personal defense inherent in the gun and especially the bullets. Lovejoy carried a gun, possibly for self-defense. But it was not designed and intended for that purpose.

The two phrases can be used interchangeably. In general self-defense is incidental, a gun fired at someone suddenly become a threat; and personal defense is purposive, a gun is owned and loaded in anticipation of being fired at a threatening person.

Another episode in the remaking of self-defense into personal defense were the remarks of New York City Police Commissioner John Enright at a conference in May, 1922.[354] The Commissioner decried private citizens arming themselves against criminals who might mug them or invade their homes. Most people are such bad shots, they could not hit the broad side of a barn, he was quoted as saying. The reporter of his words was quick to deplore this characterization. Many private citizens are former soldiers with firearms training. They have no difficulty hitting an advancing enemy or criminal entering the private residence. The site of defense is the home, but the introduction of individual military experience into home defense marks a trend toward personalization. The ammunition is not specified, only the war.

Another factor in the trend toward personal defense with hollow points is visible in stories like the one that follows.[355]

> Bradford, Pa., Dec. 24.—At Olive-dale, a small hamlet near here, yesterday afternoon Mrs. Edward Hudick shot John Ryan dead in defense of her honor. Ryan entered the woman's house during her husband's absence. She saw him coming and fearing trouble got a pistol out of a drawer and held it under her apron. When he attacked her she fired, killing Ryan instantly. The coroner's jury returned a verdict of justifiable homicide.

---

[354] Enright on Self Defense, *The New York Times*, May 19, 1922
[355] Man's Life for Honor, *Guthrie Daily Leader* [Oklahoma], December 24, 1902: 1 [Associated Press story]

This honor killing was spontaneous and defensive. Readers would have understood that John Ryan merely by entering the house and advancing toward Mrs. Budick (her own first name is not given) during her husband's absence intended to have his way with her. Not seeing the weapon beneath her apron, Ryan approached close enough to be killed with a single shot. The type of gun, caliber and bullet design were not reported. Mrs. Budick evidently knew where to find the gun quickly and that it was loaded.

Had Ryan succeeded in his plan the burden would have been on Mrs. Budick, which she and her husband anticipated by planting the gun accessibly in the first place. A woman defends her own honor, which is equivalent to her reputation and her ability to play her role in society of a small hamlet or the nation as a whole. The coroner's jury, charged with ruling on the manner of death, ruled that it was a homicide but within the bounds of legal action under the common law. Mrs. Budick was not defending her home or property; she was defending herself as a physical and social person. That an event in rural Pennsylvania could be taken up by the Associated Press and be featured on the front page of a newspaper published in the new state of Oklahoma (1907) conveys the wide intelligibility of the honor ethos.

Mr. Budick was not present to defend his house from the intruder, but he would not be charged with homicide had he been present and used deadly force to stop the man. The gun halts the invasion of property and person with one single shot. If there were any danger of the bullet passing through the assailant as the colonial troops believed the small calibers would through ghazis and the like, a bullet with relatively greater stopping power would certainly be appealing. As of 1909, however, the danger of the unstoppable assailant was not perceived in self-defense firings. Only in warfare and then as an accusation were hollow point bullets named. Bullet advertisements in newspapers and magazines offered them as an improvement of the "solid" .22 rounds used in hunting.[356]

Reports of shootings labeled as "self-defense" or "personal defense" during the ensuing decades do not indicate the type of bullet. If they name anything at all it is the caliber. The type of gun (handgun, rifle) rarely

---

[356] Appeared with variations in newspapers printed 1910-11 in Utah, New Mexico, Arizona and Washington towns.

is given. This contrasts with the preparation for self-individual defense increasingly urged upon individual citizens as a response to potential home invasion. Bullets are not advertised specifically for personal defense until the end of the century, but before that instances of instruction in preparedness for private citizens that favor certain means occasionally appear in newspapers and periodical literature.

The Queens Village Pistol and Rifle Range in Jamaica, Queens, New York was a community gun center established in a former bowling alley where local residents could receive firearms training.[357] The organizers of the Range favored shotguns over pistols in the hands of their trainees because, they stated, the average home defender did not have much experience handling guns. In the event of a sudden onrush by an intruder the defender might not be able to aim the weapon and was in danger of sending a bullet through walls and into a neighbor. The broad spray of shotgun pellets was more likely to serve as discouragement.

The military weapons training that was set against Chief Enright's dismissal of citizen defense sixty years earlier was no longer sufficiently widespread. And Mrs. Budick's preparation and presence of mind at the beginning of the century were equally out of consideration. The few who owned and could aim pistols were welcome to use the Queens Village range. Neophytes, whom the range owners planned to attract, would have more success with shotguns.

Another expedient prior to the advance of hollow point bullets was repeat firing, as possible with semi-automatic weapons. Chinatown businessman David Tse fired 18 shots into Andy Liang, who Tse said was trying to extort money from him.[358] Tse paused twice to reload his .38 revolver. At the conclusion of his trial in 1991 three years after the shooting, Tse was acquitted by jurors as acting in self-defense. He was on "automatic pilot," explained jurors interviewed afterward. The shooter was exonerated by the prodigality of his defensive response. In this environment of aggressive fear the single bullet with assured stopping power has a suggestive edge. Bullets are seldom named in accounts of shootings judged to be self-defense but shootings can make a case for an immediately effective bullet as did the circumstances of colonial armed encounters, prison government or urban policing.

The frequency of the appearance of the phrase "hollow point bullet" in English language texts from the 1890's to 2000 traces a steady increase compared with "dum dum bullet" and "soft nose bullet."[359] Like the other two phrases "hollow point bullet" enjoys a brief initial rise from zero to modest frequency between 1895 and 1903, the result of initial use in colonial wars. After a hiatus it appears again in 1908. Its frequency rises and plunges corresponding to the beginning and end of the World Wars, before starting a steady rise to 2000, at which point it exceeds the other two phrases. "Dum dum bullet" and "soft nose bullet" had their own careers reviewed earlier; references to them after 1960 are mostly historical, while references to hollow point bullets are to current use. "Hollow point bullet" also best corresponds to the rise in frequency of "self-defense" and "personal defense."

Many other n-grams certainly can be found associated with hollow point bullet. The cumulative rise of the phrase after a weak beginning compared to the other two bullet phrases implies that its context of usage is

---

[357] Geist (1982)

[358] Fraser (1991)

[359] Study performed by entering these phrases into the Google N-Gram Viewer for English. The writings searched for the phrase are the corpus of books and periodicals (few newspapers) scanned into Google's online database.

proliferating. The attachment of hollow nose bullets to personal defense in addition to historical references aggregates greater frequency numbers.

Does the appearance of hollow point bullets in reports of shootings correspond to the higher frequency of the n-gram? Or are the bullets increasingly being urged as a means of personal defense without actually being used?

Bullets still are infrequently named in cases of shootings reported in newspapers. Where they are named they appear to be linked to an expectation of personal defense on the part of the shooter. The publicized adoption of hollow point bullets by police departments and government agencies is both a sign of and an influence on the personal defense hollow points.

The trial of Philadelphia journalist Mumia Abu-Jamal for the 1981 murder of police officer David Faulkner generated considerable analysis of bullets and ballistics.[360] It was stated in some reports that the projectile that killed Faulkner was a hollow point, but in fact it was a .38 Special +P bullet, a hollow *base* round uniquely produced by the Federal firearms company. Less attention was given to the hollow point bullet Faulkner fired into Mumia, who survived the shooting and in 2012 had his death sentence reduced to life imprisonment. For a police officer to be carrying hollow point bullets was not unusual, though the expected result of their use against an alleged assailant was different from the outcome in this case.

Three years later Bernhard Goetz asserted self-defense after shooting four young men who he claimed were about to mug him on a subway car. Goetz fired five shots from a .38 caliber Smith and Wesson revolver. There was no assertion or admission of hollow point bullets. Goetz was charged with attempted murder but at the end of his trial was convicted only of illegal possession of a firearm, the gun not having been licensed. This much debated shooting spurred a National Rifle Association campaign to guarantee the right to carry concealed weapons (licensed) in public places.

A point to which a line could be drawn toward the greater association of hollow points with self-defense was the third trial in 2012 of Dale Hurd, who was accused of murdering his wife with a gunshot to the heart in 1993.[361] Hurd's defense asserted that he was alone with his wife in a room of their house showing her how to defend herself with a handgun when the gun accidentally went off. On the television were scenes of the riots that followed the acquittal of Los Angeles police officers accused of beating Rodney King after a high speed chase in 1991. The bullet that entered Bea Hurd's body was a hollow point, determined from the character of the wound and the other bullets in the gun. Dale Hurd had asked a work colleague about hollow point bullets and how loud a sound they make when fired.

This and other testimony helped convict Hurd of voluntarily shooting his wife with a bullet he believed would not leave a chance of failure. The gun was loaded under the pretense of self-defense against rioters who had to be stopped.

In the same year of Hurd's final trial two other incidents occurred that underlined the growing presence of hollow point bullets under the same self-defense pretense that Hurd use to explain handling a hollow point bullet loaded gun in his wife's presence.

---

[360] Peheim (2007)
[361] Leonard (2012)

In July, 2012 Christopher Joyce, an economics student at University of Florida-Gainesville, was awakened from sleep in his dormitory bed by a sharp pain in his leg. He noticed a small circular wound in his leg, and a hole in the wall of the room corresponding to his leg's position.[362] Later a surgeon told him that shrapnel from the bullet that struck him was too close to vital blood vessels to be removed. Security guard Jimmy Dixon, who was in the building at the time of the shooting, had 9mm hollow point bullets in his service revolver rather than the 9mm bullets the security firm said he was issued. The standard 9mm bullet would have passed through Joyce; the hollow point mushroomed and broke up within his leg. Joyce had fallen asleep sitting up. If he had been lying down when the bullet struck, a doctor speculated, the injury would have been much graver. An attorney filed a lawsuit in civil court on Joyce's behalf for damages.

The bullet had proceeded as a hollow point is advertised to do, passing through intervening barriers to expand inside the limb of the person it struck. It had not been inhibited by clogging with wall matter and so must have been a hollow point designed not to lose kinetic energy until it hit skin and deformed. Stated security firm policy was incompatible with the practice of their employees, or at least of this one employee, who armed himself for personal defense and then simply fired the bullet into a wall.

After George Zimmerman shot and killed Trayvon Martin during a disputed encounter on February 26, 2012, news reports explained why Zimmerman was carrying a pistol loaded with hollow point bullets.[363] A pit bull in the neighborhood had run loose several times, and young black males were suspected of committing robberies. As chief of the neighborhood watch in the Sanford, Florida gated community, Zimmerman patrolled the streets looking for possible burglars and on the night of the shooting he reported to police that he saw a man who turned out to be Martin taking short cuts among the houses and looking in windows. After the shooting Zimmerman was questioned by police and released on the grounds that he had fired in self-defense. Witness testimony and a reevaluation of the evidence caused him to be arrested later on the charge of second-degree murder, and he now awaits trial.

Zimmerman was licensed to carry a handgun at the time of the shooting though the neighborhood watch committee and the local police did not encourage him to do so while he was on his rounds. The fact that it was loaded with hollow point bullets became public knowledge but did not form part of the extensive commentary on Zimmerman's preparations or the accounts of the shooting itself. The bullets were mentioned in separate reports on the light, easily concealed gun that Zimmerman carried.

Trayvon Martin's autopsy report concluded that the fatal round had made its entrance at the left chest "from intermediate range" and there was no exit wound. "Fragments of projectile recovered in pericardial sac and right pleural cavity." [364] The report did not mention that a hollow point bullet was the most likely cause of this injury, nor did any of the news reports summarizing the autopsy. They concentrated on the "intermediate range" conclusion, which was established because there were no powder burns at the site of the entrance wound.

---

[362] Alcantara (2012)

[363] Wright (2012)

[364] Office of the Medical Examiner. Florida Districts 7 & 24 (2012)

Zimmerman stated that he began carrying the gun armed with the hollow points after an escaped pit bull cornered his wife and an animal control officer gave him that advice. That is the preferred ammunition for urban animal control, and for the type of gun Zimmerman carried, a Kel-Tec 9mm PF-9, made for personal defense.

Several narratives converged in the Martin shooting that could not be resolved by court action: race, urban crime and policing, municipal finances, personal defense laws. The hollow point bullet narrative assembled itself from the elements associated with it, and with expanding bullets in general. Apprehensive of attack by animals, Zimmerman chose to defend himself and his family with a bullet designed to stop an assailant. Whether or not Martin was an assailant and what kind of assailant is in dispute.

The gun and its bullets carried over to Zimmerman's role as a neighborhood watchman. Convinced he was defending the neighborhood against a burglar, he engaged with and shot an unarmed man. He was not behind a barricade facing determined warriors, but the structure of his weaponry had that history, especially when he removed it from concealment.

Neither of the shooters in these two incidents was a police officer; they were employed or volunteered as guards or guardians. Their actions took place in the midst of widespread adoption of hollow point bullets by police departments and other agencies with a policing role. The order for 450 million .40 caliber hollow point bullets, 90 million a year on a 5 year contract, placed by the United States Department of Homeland Security was explicable if not reassuring.[365] The Social Security Administration's 174,000 hollow point bullet order was attributed to the need for investigators of Social Security fraud to defend themselves.[366]

Among the new products Hornady, for one, announced for the 2013 season are three bullets in their Critical Duty line of hollow points: 9mm, 9mm+P and .40 S&W.[367] Their attempts to perfect a .45 hollow point were hampered by the difficulty controlling expansion when the bullet reaches the skin. The cross-section of the bullet's nose carries too much kinetic energy meeting the membrane's plane and is in danger of expanding on contact rather than usefully inside the target. The Triple Defense .410 shot shell contains a .410 unjacketed FTX slug and two .35 hand loaded shot balls all of which would act as a hollow point. The manufacturer clearly believes that hollow point design is a strong selling point.

---

[365] Lindorff (2012)
[366] Tomasky (2012)
[367] Slowik (2012)

17. The Gendered Frame

The advent of the dumdum complex and associated bullets at the end of 19th century coincided with a change in the association of women with firearms. Prior to that period women, in the United States and especially on the western frontier, regularly handled guns for practical purposes and were depicted in popular culture doing so. Women of all groups and class background used guns to hunt for subsistence and in warfare. Women could be sharpshooters defeating men in demonstrations of uncanny skill. Annie Oakley, who had been a hunter and a trapper, was a masterful female presence in the otherwise male-dominated touring Wild West Show.

Annie Oakley did not believe she had to abandon the women's role of her time in her career as an exhibition sharpshooter, and in fact she expected maintaining that role to help her career.[368] Female gunslingers and female gangsters, in fact and in fiction, did not abrogate the female role and some evidently did not want to. Bonnie Parker, on the lam robbing banks with Clyde Barrow during the 1930's, celebrated the pair's exploits in poetry she sent to newspapers, but was disconcerted when a photograph of her holding a machine gun and smoking a cigar also appeared in the press.[369]

As the frontier and the wars of previous century receded, women were more likely to take up guns strictly for self-defense and then under male tutelage. Mrs. Budick of the previous chapter held under her apron a gun concealed for the purpose of fending off a possible attacker in the absence of her husband. Bea Hurd's police officer husband in 1993 was supposedly instructing her in defensive gun use during urban riots when the gun went off. Women have hunted game and been sharpshooters during the 20th century as they had during the 19th. They were pictured and counseled to see themselves as potential victims of unrestrained, brutal males. Not suffering a fate worse than death-rape by a foreigner-called for a gun and gun skills to be readily at hand.

The structure of women's position with regard to guns in general resembled the structure of dumdums and the like for male constituencies. The intruder a woman was called on to defend herself and her home against resembled the fired-up assailant stopped by a dumdum bullet where conventional rounds failed. Both invaders were male fiends on a rampage. For women the lines of battle where the soldiers held out against the ghazis were drawn around the home and the personal space only a male with no regard for convention would breach.

Manufacturers have long made light-weight, low recoil hunting rifles marketed specifically for women. They now produce handguns designed as fashion accessories, and suggest that handguns with certain features are more appropriate for women's self-defense. This strain has emerged in gun advertising as self-defense has become a gun selling point for both men and women. The general threat has shifted from overseas barricades to the home, the workplace and the street. This shift has in turn brought along the single shot stop bullets that expand in flesh after penetrating glass or wood barriers. Gun makers seek to form a women's marketing niche within this larger category of defensive weapons. The expanding bullets come to women not

---

[368] Riley (1994: 123)
[369] Browder (2006: 12-13)

for women's purposes alone but as part of the self-defense package much as hollow points came to women's hunting rifles as to all hunting rifles.

The access to guns women have been offered for their personal protection has been extended to the domestic sphere including men and children, and to professions such as teaching with no inherent use for guns. The school and the college are now to be self-defended by gun carrying instructors.

Women assembled cartridges and cast bullets from the earliest descriptions of the process, especially under wartime conditions that required mass production. Cartridge making from the American Revolutionary War to the Civil War was a home craft practiced by women in support of male kin and community members. African slave women loaded cartridges for British army officers during the Revolutionary War in the hope of gaining their freedom. Rural African-American women could find jobs in munitions plants located away from the cities force during World War I.[370]

The cover illustration of an issue of *Harper's Magazine* dated July 20, 1861 depicts the process of cartridge loading at the Watertown, Massachusetts arsenal. The first hostilities of the Civil War had taken place, and the demand for ammunition was rapidly increasing as the war spread. Winslow Homer drew the cartridge loading in two phases. Below, men pour gunpowder through funnels into paper cones lodged in frames. Above, rows of young women place a Minié ball atop the gunpowder, and enclose the cartridge in another paper envelope which they fold closed and stack in an ammunition box. A man in uniform supervises the women's work. The women's hats and shawls hang on the wall behind the work table.

The task of scooping and pouring the gun powder was the province of the men, only a few required to turn out enough filled cartridges to occupy a larger number of women with the more time-consuming tasks of loading, folding and arranging the final product. Women came to arsenal work in numbers to replace the men called to the war, and they were compensated less than men for the same work. Though Homer's illustration implies that the men's work was more dangerous, explosions and fires at arsenals killed more women than men.[371]

A *Harper's Weekly* cover that appeared 16 years later maps the social transformation of cartridge making with mechanization. Theodore G. Davis's wood engraving of the work at the Union Metallic Cartridge Company in Bridgeport, Connecticut places the women at individual work stations where they fix the bullets in metal cartridge tubes that have been automatically loaded with gunpowder from a hopper above the turntable rack in which the cartridges circle. The men have been reduced to attendants delivering the raw materials and removing the production. The only remnant of their previous role is pouring the gunpowder into the hoppers. Charles H. Fitch's 1882 report on the manufacture of firearms and ammunition baldly states that it is mainly women who operate the bullet press machines.[372]

A succession of photographs of this process up to the munitions factories of World War II shows the same arrangement with a greater integration of the separate processes into the machine and woman operator combination lined tightly with identical others. Despite women's continued predominance in ammunition

---

[370] Denman and Inniss (1999: 192)
[371] Slavicek (2009: 74-75)
[372] Fitch (1882: 35)

manufacture, the rhetoric of each war treats women as replacing men in the work for the duration of the war.[373]

This is the rhetoric employed with unintentional irony in a satirical drawing by Thomas Theodor Heine printed in the December 1, 1914 issue of the German magazine *Simplicissimus*. The purpose of the drawing is to associate women making cartridges at home during wartime with house work and with dumdum making as in the Damarli field factory of the British Sudan campaign (Illustration 21). Entitled "A Picture from English Family life" ("Ein Bild aus dem Englischen Familienleben"), the caption explains "Thus in England in household circles one is pressed to help the soldiers. The ladies fill in making dum-dum bullets." Around the parlor table and in chairs nearby, where sewing circles usually meet, elderly women and a young girl are using household implements such as pliers, vises and bread cutters to clip the tips of bullets. One woman slices at the bullet with scissors; another presses the bullet with her extracted dentures. On the floor a pug is standing up and biting off the end of a bullet with his own teeth. The ladies are filling in for the men who are off at war, but they are English ladies and so they fill in making dumdums with their household implements. German ladies may be making cartridges for the cavalryman shown attentive with his horse on the cover of the magazine. English ladies have a more perverse use for their skills.

This anti-English satire marks a threshold in the history of bullets apart from its parochial intent. It encompasses many of the changing bullet elements-national and cultural boundaries crossed, techniques of modification, inherent malice-and it highlights an element becoming explicit about changing bullets as it had about bullets in general: gender. Gunpowder impelled projectiles long were a male province, to make and to use. The development of bullet cartridges calling for packaging gave women an opportunity to contribute to male success abroad. The division between male bullet users and female bullet makers was only intensified by the industrialization of the enterprise.

Most of the bullets used by men were made by women: the bullet made to change, to become more penetrating when it reached flesh, was a response to that hidden history. Initially men in the field made dumdums from bullets produced by women in factories at home in an effort to reclaim their command of bullet shape and production. The satirical drawing postulated that even dumdums could be given over to women. By this time, the World War I dumdum complex was active and the give and take of the bullets had become autonomous. Everyone used them, claimed they didn't and that others did. Everyone believed they were especially effective at stopping the unstoppable assailant and yet that they were like every other bullet type.

Self-defense, shifting the barricades of the dumdum's creation to the home and then to the individual was one solution to the ambiguity of the bullets, their slipping gender orientation and bilateral direction. The former clear cut enemies no longer were overseas; they had come home. Anyone might be a bullet user against the enemy who might be anyone. Women were recruited to this defensive position and supplied with bullets they didn't make. A reunited community of users forms the whole against a common enemy similarly armed. Men and women, Christians and Turks, police and criminals, all are firing dumdums against their (our) assailants. This is what the bullets say.

---

[373] Thom (1995); Wightman (1995)

60. Winslow Homer cover for Harper's Weekly, July 20, 1861.

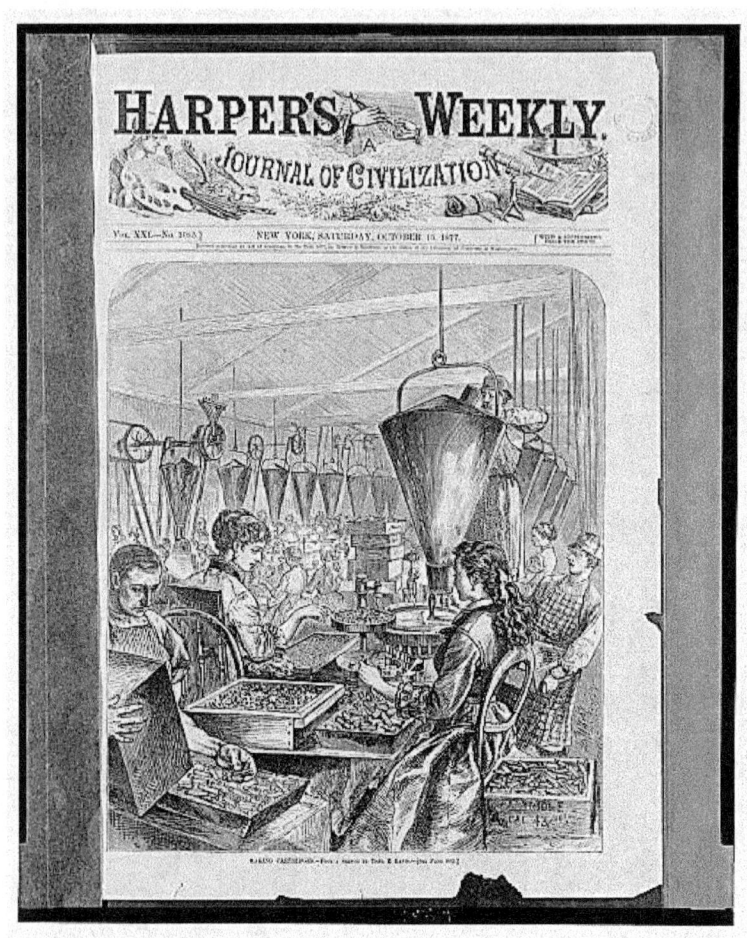

61. *Theodore G. Davis cover for Harper's Weekly, October 15, 1877*

Ein Bild aus dem englischen Familienleben

(Zb. Th. Heine)

Auch in England ist man im häuslichen Kreise eifrig bestrebt, den Soldaten zu helfen. Die Damen wetteifern in der Herstellung von Dum-Dum-Geschossen.

— 474 —

62. Thomas Theodor Heine illustration for Simplicissimus, December 1, 1914

References

Ágoston, Gábor. 2005. *Guns for the Sultan: Military Power and the Weapons Industry in the Ottoman Empire*. Cambridge: Cambridge University Press.

Alcantara, Chris. 2012. UF Student, Shot in Leg in July, Files Charges, *The Independent Florida Alligator* [online] November 16

Ambrose, Stephen. 1994. *D-Day: June 6, 1944- the Climactic Battle of World War II*. New York: Simon and Schuster.

Ameringer, Oscar. 1940. *If You Don't Weaken: The Autobiography of Oscar Ameringer*. New York: Henry Holt.

[Appleton, D.] 1899. *The Annual Encyclopedia and Register of Important Events for 1899*. New York: D. Appleton and Co.

Awad, Dina and Hazem Jamjoum. 2008. Diaries: Live from Palestine, "I Do Not Struggle Alone," *The Electronic Intifada* [online] July 15, 2008.

Ayoob, Massad. 2012. *Gun Digest's Concealed Carry Gun Ammo eShort*. Gun Digest Books [online]

Barnum, Phineas Taylor. 1866. *The Humbugs of the World*. New York: Carleton Publishers.

Barrett, David. 2011. Norway Massacre: British Traders Helped Supply Breivik's Arsenal of Weapons, *The Telegraph* [online] July 30

Beevor, Antony. 2009. *D-Day: The Battle for Normandy*. New York: Viking.

Belich, James. 2001. *Paradise Reforged: A History of New Zealanders from the 1880's to the Year 2000*. Honolulu: University of Hawai'i Press.

Bennett, Ernest Nathaniel. 1899. *The Downfall of the Dervishes; Or, The Avenging of Gordon*. London: Methuen and Co.

Bennett, Ernest Nathaniel. 1900. *With Methuen's Company on an Ambulance Train*. London: Swan Sonnenschein and Co.

Beyer, Henry. 1894. On Gunshot Injuries Produced by the New Projectiles of Small Caliber, *Proceedings of the U.S. Naval Institute* 20: 149-67.

Bingham, W.H., trans. 1905. After Mukden: A Russian Verdict on Russian Failures, *Journal of the Royal United Service Institution* 49: 686-95.

Breivik, Anders Behring (Andrew Berwick). 2011. *2083: A European Declaration of Independence*. London.

Brewster, David, ed. 1832. *The Edinburgh Encyclopedia, v. 18*. Philadelphia: John Parker.

Browder, Laura. 2006. *Her Best Shot: Women and Guns in America*. Chapel Hill: University of North Carolina Press.

Brown, David. 2012. *Safari 101 Hunting Africa: The Ultimate Adventure.* New York: Morgan James Publishing.

Calabi, Silvio. 2011. Ernest Hemingway on Safari: The Game and the Guns, 85-121 IN *Hemingway in Africa,* edited by Miriam B. Mandel. Rochester, NY: Camden House.

Carless, Albert. 1898. General Surgery, *The Practitioner* 61: 76-88.

Chagnon, Napoleon. 2013. *Noble Savages: My Life Among Two Dangerous Tribes-the Yanomamo and the Anthropologists.* New York: Simon and Schuster.

Chambers, G. Paul. 2010. *Head Shot: The Science Behind the Kennedy Assassination.* Amherst, NY: Prometheus Books.

Chet, Guy. 2003. *Conquering the American Wilderness: The Triumph of European Warfare in the Colonial Northeast Wilderness.* Amherst, Massachusetts: University of Massachusetts Press.

Chevalier, Thomas. 1806. *A Treatise of Gun-Shot Wounds.* London: Samuel Bagster.

Churchill, Winston. 1901. *The Story of the Malakand Field Force: An Episode of Frontier War.* London: Longmans, Green and Co.

Cirillo, Vincent. 2004. *Bullets and Bacilli: The Spanish-American War and Military Medicine.* Piscataway: Rutgers University Press.

Cleveland, Harold Irwin. 1900. *Massacres of Christians by Heathen Chinese and Horrors of the Boxers.*

Cleveland, William H. and Martin Bunton. 2009. *A History of the Modern Middle East.* Boulder: Westview Press.

Corporal of the Marines. 1900. Besieged in Pekin, *The South African Magazine* 2: 205-60.

Cottar, Charles. 1915. Some Things Guns Will and Will not Do and Some of the Reasons Why, *Arms and the Man* 59, 1: 10-12.

Crosby, Alfred W. 2002. *Throwing Fire: Projectile Technology Throughout History.* Cambridge: Cambridge University Press.

Crossman, Edward C. 1916a. The Dum-Dum Yarns, *Journal of the Services Institution of the United States* 58: 129-35.

Crossman, Edward C. 1916b. "Dum-Dum Bullets," *The Scientific American War Book,* 153-55.

Crutcher, Howard. 1906. Further Observations on the Use of the Soft-Point Bullet, *The Illinois Medical Journal* 8: 499-500.

Dash, Mike. 2009. *The First Family: Terror, Extortion, Revenge, Murder and the Birth of the American Mafia.* Doubleday Canada.

Dash, Mike. 2012. The Blazing Career and Mysterious Death of "the Swedish Meteor," Past Imperfect, Smithsonian.com, September 17.

Davis, Gwillym. 1897. The Effects of Small Calibre Bullets as Used in Military Arms, *Annals of Surgery* 25,1: 36-50.

Davis, Theodore. 1869. The Buffalo Range, *Harper's Magazine* 38: 147-63.

Denman, Joan B and Leslie Baham Inniss. 1999. No War Without Women: Defense Industries, 187-99 IN *Gender Camouflage: Women and the U.S. Military.* New York: New York University Press.

Dent, Clinton. 1900a. The War in South Africa, *British Medical Journal* 2051: 968-74.

Dent, Clinton. 1900b. Surgical Notes from the Military Hospitals of South Africa, *British Medical Journal* 2043: 471-73.

Devereux, George. 2000. Normal and Abnormal, 213-89 IN *Cultural Psychiatry and Medical Anthropology,* edited by Roland Littlewood and Simon Dein. London: The Athlone Press.

Díaz Cárdenas, León. 1989. *Cananea: Primer brote del sindicalismo en México.* México: Centro de Estudios Hispanicos del Movimiento Obrero Mexicano.

Dixie, E.A. 1921. An Old Machine-Gun Patent, *American Machinist* 54,4: 146.

Dobson, Austin. 1896. *Eighteenth-Century Vignettes: Third Series.* New York: Dodd, Mead.

Dockery, Kevin. 2007. *Future Weapons.* New York: Penguin.

Duff, Adrian C. 1914. Eye Pictures of Vera Cruz by the Camerman, *The Princeton Union* (Minnesota) June 18: 2.

Dutertre-Dulévièleuse, Émile. 1916. *Balles dum-dum, balles explosibles, balles explosibles Autrichiennes, 1914-1916.* Paris: A. Maloine et Fils.

Edgerton, Robert B. 1997. *Warriors of the Rising Sun: A History of the Japanese Military.* New York: W.W. Norton.

Edwards, Frank. 2004. *The Gaysh: A History of the Aden Protectorate Levies 1924-61 and the Regular Army of South Arabia, 1961-67.* Solihull: West Midlands Press.

Ellis, Havelock. 1915. *The Criminal.* London: The Walter Scott Publishing Company.

Ellis, John. 1975. *The Social History of the Machine Gun.* New York: Pantheon.

Ernle, R.E.P., ed. Ireland. *The Quarterly Review* 236: 155-72.

Escobar, Edward J. 1999. *Race, Police and the Making of a Political Identity: Mexican-Americans and the Los Angeles Police Department.* Berkeley: University of California Press.

Esposito, Barbara and Lee Wood. 1982. *Prison Slavery.* Washington, D.C.: Committee to Abolish Prison Slavery.

Eyffinger, Arthur. 1999. *The 1899 Hague Peace Conference.* The Hague: Kluwer Law International.

Fackler, Martin. 1997. Book Review: Marshall, E.P., Sanow, E.J., Street Stopper, *Wound Ballistics Review* 3,1: 26-31.[online reprint]

Farago, Roberto. 2011. Norwegian Spree Killer Anders Behring Breivik Used Ruger M-14, *The Truth About Guns* [online] July 25.

Ferriman, Anabel. 2002. Palestinian Territories Face Huge Burden of Disability, *British Medical Journal* 324, 7333: 320.

Feuer, Alan. 2012. Fatal Shootings Raise Questions About Police Training, *The New York Times* December 9: 38-39.

Ffoulkes, Charles and Lord Cottesloe. 2011 (1937). *The Gun-Founders of England.* Cambridge: Cambridge University Press.

Field, Michael. 2006. *Black Saturday: New Zealand's Tragic Blunders in Samoa.* Reed Publishers.

Firearms Tactical Institute. 1998. Tactical Briefs #2, March 1 [online]

Fletcher, R. 1891. *The New School of Criminal Anthropology: An Address Delivered before the Anthropological Society of Washington.* Washington: Judd and Detweiler.

Foltz, Fredk. 1901. Penetration of Soft-Nosed Bullets in Flesh, Bone, etc. [letter], *Naval Institute Proceedings* 27: 838.

Fraser, C. Gerald. 1991. 18-Shot Killing is Ruled by Jury as Self Defense, *The New York Times* July 15

Fulton, Robert. 2011. *Honor for the Flag: The Battle of Bud Dajo-1906 and the Moro Massacre.* Bend, Oregon: Tumalo Creek Press.

Fyfe, Herbert. 1900. The Röntgen Rays in Warfare, *The Strand Magazine* 19: 538-44.

Garrison, Fielding. 1918. *An Introduction to the History of Medicine.* Philadelphia: W.B. Saunders Company.

Geary, Grattan. 1886. *Burmah, After the Conquest...*London: Sampson Low, Marston, Searle and Rivington.

Geist, William G. 1982. Students Learn Self-Defense at a Rifle Range, *The New York Times* April 2.

Gerard, Jeremy. 1989a. PBS Investigating Financing of 'Days of Rage', *The New York Times* August 31.

Gerard, Jeremy. 1989b. PBS Executive Resigns in 'Days of Rage' Dispute, *The New York Times* September 9.

Gérard, Jules. 1856. *Lion Hunting and Sporting in Algeria.* London: Addey and Co.

Gettleman, Jeffrey. 2012. Elephants Dying in an Epic Frenzy as Ivory Fuels Wars and Profits, *New York Times,* September 3: 1+

Fitch, Charles. 1882. *Extra Census Bulletin: Report on the Manufacture of Fire-arms and Munitions.* Washington: Government Printing Office.

Gilbert, Martin. 2005. *The Routledge Atlas of the Arab-Israeli Conflict.* Abingdon: Routledge.

Gleeson, James. 1962. *Bloody Sunday: How Michael Collins' Agents Assassinated Britain's Secret Service in Dublin on Nov. 21, 1920.* London: Peter Davies.

Great Britain. Parliament. House of Commons. 1901. Minutes of Evidence Taken Before the Committee of Public Accounts, 15 May 1901 Compensation to Contractors, *Papers by Command, v. 5.* London: His Majesty's Stationery Office.

Great Britain. Parliament. House of Commons. 1908. Appendix-Report of the Trial of Natives for the Murder of Mr. Stainback, 221-32 IN *Further Correspondence Relative to Native Affairs in Natal, In Continuation of [Cd. 3247] and [Cd. 3563].* London: His Majesty's Stationery Office.

Great Britain. Patent Office. 1902. *Patents for Inventions: Abridgement, Class 9: Ammunition, etc.* London: His Majesty's Stationery Office.

Great Britain. War Office. Woolwich Ordnance College. 1902. *Treatise on Ammunition.* London: His Majesty's Stationery Office.

Grogan, Ewart Scott and Arthur H. Sharp. 1902. *From Cape to Cairo: The First Traverse of Africa from South to North.* London: Hurst and Blackett, Ltd.

Gusfield, Jeffrey. 1928. *Deadly Valentines: The Story of Capone's Henchman "Machine Gun" Jack McGurn.* Chicago: Chicago Review Press.

Haami, Bradford. 2004. *Putea Whakairo: Maori and the Written Word.* Wellington: Huia Press.

Hackin, Charles. 1923. *Guide-Catalogue du Musée Guimet. Les Collections Bouddhiques.* Paris: G. van Oest and Co.

Halileh, Samia, et al. 2002. The Impact of the Intifada on the Health of a Nation, *Medicine, Conflict and Survival* 18: 239-48.

Hamilton, J.B. 1898. The Evolution of the Dum-Dum Bullet, *British Medical Journal* 1, 1950: 1250-51.

Hårdh, Robert. 2012. Swedish police should stop using expanding bullets, *Civil Defenders,* April 23 [online]

Hart, John Mason. 1987. *Revolutionary Mexico: The Coming and Process of the Mexican Revolution.* Berkeley: University of California Press.

Hensman, Howard. 1882. *Afghan War of 1879-80.* London: W.H. Allen and Co.

Hesketh-Pritchard, Hesketh. 1920. *Sniping in France.* London: Hutchinson.

Higginson, John. 2009. Fears over use of "dum dum" bullets by police, *Metro* [online]

Holmes, Richard. 2011. *Sahib: The British Soldier in India, 1750-1914.* London: HarperCollinsUK.

Holub, Emil. 1975. *Emil Holub's Travels North of the Zambezi, 1885-86,* translated by Cynthia Johns. Manchester: Manchester University Press.

Horovitz, David. 2012. *Shalom, Friend: The Life and Legacy of Yitzhak Rabin.* Newmarket Press.

Hunter, John. 1828. *A Treatise on the Blood, Inflammation and Gun-shot Wounds.* London: Shorewood, Gilbert and Piper.

Hooper, W.W. and V.S.G. Western. 1887. *Tiger Shooting in India.* J.A. Lugard.

Hopkinson, Michael. 2002. *The Irish War of Independence.* McGill-Queens University Press.

Hurt, Robin. 2006. Foreword IN *Dangerous-Game Rifles* by Terry Wieland. Camden: Countryside Press.

Isenberg, Andrew. 2000. *The American Bison: An Environmental History, 1750-1920.* Cambridge: Cambridge University Press.

James, Lawrence. 1998. *Raj: The Making and Unmaking of British India.* London: Macmillan.

Jamjoum, Hazem. 2008. Ramallah Commemorates the Ongoing Nakba, *The Electronic Intifada* [online] May 28, 2008.

Jones, David E. 2007. *Poison Arrows: Native American Indian Hunting and Warfare.* Austin: University of Texas Press.

Katz, Friedrich. 1998. *The Life and Times of Pancho Villa.* Stanford: Stanford University Press.

Kirk, Bryan. n.d. The 101 California Street Killings and Gun Control Litigation: *Merrill v. Navergar, Inc.* http://www.law.berkeley.edu/sugarman/Merrill-1.doc

Kaye, Brian. 1996. *Golf Balls, Boomerangs and Asteroids: The Impact of Missiles on Society.dddd*

Kneubuhl, Beat. 2011. Wound Ballistics of Bullets and Fragments, 163-252 IN *Wound Ballistics: Basics and Applications,* edited by Beat Kneubuhl. Berlin: Springer Verlag.

Knight, B. 1982. Explosive Bullets: A New Hazard for Doctors, *British Medical Journal* 284: 768-69.

Knight, Ian. 2009. *Maori Fortifications.* Oxford: Osprey.

Knox, Ernest Blake. 1902. *Hooker's Campaign with the Natal Field Force of 1900.* London: R. Brimley Johnson.

Körtner, A. 1907. Firing Tests and Experiences with the Small-Caliber Rifle, translated by William Lassiter, *The Military Surgeon* 20, 5: 396-406.

Koskimaki, George. 2006. *D-Day with the Screaming Eagles.* New York: Presidio Press.

Kronabel, Dieter and Ralf Vollmuth. 2011. Ein Rezept aus dem Späten Mittelalter zur Behandlung von Schusswunden, *Wehrmedezin und Wehrpharmazie* [online]

Kuei, Chung-shu. 1932. *Symposium on Japan's Undeclared War in China.* Chinese Chamber of Commerce.

Kuitenbrouwer, Vincent. 2012. *War of Words: Dutch Pro-Boer Propaganda and the South African War (1899-1902).* Amsterdam: Amsterdam University Press.

Lagarde, Louis. 1916. *Gun Shot Injuries: How they are Inflicted, their Complications and Treatment.* New York: William Wood and Company.

Lahav, Pnina. 1993. The Press and National Security, 173-96 IN *National Security and Democracy in Israel,* edited by Avner Yaniv. Boulder: Lynne Rienner.

Lash, Irving. 1958, We're Not a Gadget Shop, Says the Patent Office, *Popular Mechanics* September: 11; 14.

Latham, John. 1866. Early Breech Loaders, *Royal United Service Institute Journal* 9: 88-106.

Le Wald, Leon. 1907. A Case of Multiple Gun Shot Wounds, *The Military Surgeon* 20: 192-93.

Lee, David. 2006. *Up Close and Personal: The Reality of Close Quarter Fighting in World War II.* London: Greenhill.

Lehmann, Hans. 1911. *Die Kinematographie: Ihren Grundlagen und Ihre Anwendungen.* Leipzig: B. Treubner.

Lidin, Olof G. 2002. *Tanegashima: The Arrival of Europe in Japan* (Nordic Institute of Asian Studies Monograph Series No. 90). Copehagen: NIAS Press.

Lindorff, David. 2012. The Potential Killers All Around Us, *This Can't Be Happening* April 10 [online]

Lino, Art. 1944. The Horrors of War-Told By a Marine who Lived through Hell, *The St. Petersburg Times* February 9: 1.

London, Jack. 1914a. With Funston's Men, *Collier's Magazine* 53 (May 23): 9-10; 26.

London, Jack. 1914b. Mexico's Army and Ours, *Collier's Magazine* 53 (May 30): 5-7.

Long, Robert Crozier. 1909. This Man Has Solved Problem of Photographing the Invisible, *Deseret Evening News* September 18: 1.

Longmore, Thomas. 1862. *A Treatise of Gunshot Wounds.* Philadelphia: J.B. Lippincott.

Longmore, Thomas. 1877. *Gunshot Injuries.* London: Longman, Green and Co

Lower, Wendy. n.d. The Holocaust and Colonialism in Ukraine: A Case Study of the Generalbezirk in Zhytomyr, Ukraine, 1941-44. [online]

Lower, Wendy. 2005. The *"reibungslose"* Holocaust: The German Military and Civilian Implementation of the "Final Solution" in Ukraine, 1941-44, 236-59 IN *Networks of Nazi Persecution: Bureaucracy, Business and the Organization of the Holocaust,* edited by Gerald D. Feldman and Wolfgang Seibel. Berghahn.

Lowry, Heath. 2003. *The Nature of the Early Ottoman State.* Albany: State University of New York Press.

MacGavin, Drummond. 1906. The Slaughter of the Hippopotami, *San Francisco Call* January 21: Literary Section 1.

Mackay, Charles. 1856. *Memoir of Extraordinary Popular Delusions and the Madness of Crowds,* v. 1. London: G. Routledge & Co.

Macleod, G.H.B. 1858. *Notes on the Surgery of the War in the Crimea.* London: John Churchill.

Mallinckrodt, Edward. 1922. Hunting and Photographing the Brown Bear of Alaska, *Outing* 82, 2: 51-58.

McNelly, R.W., trans. 1914. The Tragic Days of Vera Cruz (from *La Opinion* (Vera Cruz), April 23, 1914), *Proceedings of the U.S. Naval Institute* 40: 741-58.

Makins, G.H. 1900. A Note on the So-Called "Poisoned Bullet," *British Medical Journal* 2, 2068: 444

Makins, G.H. 1901. *Surgical Experiences in South Africa.* London: Smith, Elder and Company.

Makins, G.H. 1919. *On Gunshot Wounds to the Blood Vessels Founded on Experience Gained in France During the Great War, 1914-1918.* New York: William Wood and Company.

Maleisa, Malama. 1987. *The Making of Modern Samoa: Traditional Authority and Colonial Administration in the Making of Modern Samoa.* Suva, Fiji: Institute of Pacific Studies.

Manchester, William. 1983. *The Last Lion, Vol. I: Winston Churchill, Visions of Glory, 1874-1932.* New York: Little, Brown and Co.

Manchester, William and Paul Reid. 2012. *The Last Lion Vol. III: Winston Spencer Churchill Defender of the Realm, 1940-1965.* New York: Little, Brown and Co.

Marks, Alfred. 1902. Bullets, Expansive, Explosive and Poisoned, *The Westminster Review* 157: 619-35.

Marks, Shula. 1970. *Reluctant Rebellion: An Assessment of the 1906-08 Disturbances in Natal.* Oxford: Clarendon Press.

Marre, Francis. 1916. *Les armes déloyables des Allemands.* Paris and Barcelona: Bloud and Gay.

Marshall, Evan and Edwin Sanow. 1992. *Handgun Stopping Power: The Definitive Study.* Boulder, Colorado: Paladin Press.

Marshall, Evan and Edwin Sanow. 1996. *Street Stopper: The Latest Handgun Stopping Power Street Results.* Boulder, Colorado: Paladin Press.

Marshall, Evan and Edwin Sanow. 2001. *Stopping Power: A Practical Analysis of the Latest Handgun Ammunition.* Boulder, Colorado: Paladin Press.

Massachusetts. Superior Court. 1903. *The Official Report of the Trial of John C. Best for Murder.* Boston: Wright and Potter.

Maxwell, Charles H. 1906. The Witchcraft Delusion in South Africa, *The Missionary Herald at Home and Abroad* 102: 588-92.

McKnight, Gerald. 2005. *Breach of Trust: How the Warren Commission Failed the Nation and Why.* University of Kansas Press.

McMahon, Sean. 2001. *Rebel Ireland: Easter Rising to Civil War*. Cork: Mercier Press

McManus, John C. 2004. *The Americans at D-Day: The American Experience at the Normandy Invasion*. New York: Tom Doherty Associates.

Middleton, Drew. 1943. Prisoners Flock to British Tanks, Drawn to Armor as if by Magnet, *The New York Times* May 13, 1943: 1;4.

Millais, J. M. 1919. *The Life of Frederick Courtenay Selous, D.S.O.* London: Longman, Greens and Co.

Mitchell, Hilary and John. 2004. *Te Tau Ihu o te Waka: A History of the Maori of Nelson and Marlborough, v. II.* Wellington: Huia Press.

Mogelson, Luke. 2012. Najiba, *The New York Times Magazine* December 30: 22-23.

Montano, Joseph. 1886. *Voyages aux Philippines et en Malaisie.* Paris: Librairie Hachette.

de Montigny, Alan Kelso. 1953. *International Anthropological and Linguistic Review* 1

Mordecai, Alfred. 1861. *Report of the Military Commission to Europe in 1855 and 1856.* Washington: G.W. Bowman.

Morris, Charles. 1873. The Dominion of the Savage, *To-Day: the Popular Illustrated Magazine* 1,28: 534.

Morrison, George Ernest. 1986. *The Correspondence of G.E. Morrison, 1895-1912.* 2 v. Cambridge: Cambridge University Press.

Mylès, Henri. 1913. *Instantanés d'Extrême-Orient.* Paris: E. Sansot.

Naidu, Vijay. 1993. The Path to Independence, 126-44 IN *Culture Contact in the Pacific: Essays on Contact, Encounter and Response.* Cambridge; Cambridge University Press.

Najita, Susan Y. 2006. *Decolonizing Cultures in the Pacific: Finding History and Trauma in Contemporary Fiction.* New York: Routledge.

Needham, Joseph. 1981. *Science in Traditional China: A Comparative Perspective.* Cambridge: Harvard University Press.

Needham, Joseph. 1987. The Fire Lance, Ancestor of All Gun-Barrels, 295-334 IN *Chinese Ideas about Nature and Society: Studies on Honor of Dirk Bodde*, edited by Charles Le Blanc and Susan Blader. Hong Kong: Hong Kong University Press.

New Zealand. Parliament. House of Representatives. 1930. Coroner's Finding in the Inquest Respecting the Fatalities in Western Samoa, *Appendix to the Journals of the House of Representatives*, A-4b.

Nicolle, David. 1937. *The Italian Invasion of Ethiopia, 1935-36.* Oxford: Osprey Publishing.

Nimier, H. and Ed. Laval. 1899. *Les Projectiles des Armes de Guerre-Leur Action Vulnérante.* Paris: Felix Alcan.

Nugent, Daniel. 1993. *Spent Cartridges of Revolution: An Anthropological History of Namiquipa.* Chicago: University of Chicago Press.

Office of the Medical Examiner. Florida Districts 7 & 24. 2012. *Medical Examiner Report-Trayvon Martin.*

Ogden, H.J. 1901. *The War Against the Dutch Republic in South Africa: Its Origin, Progress and Results.* Manchester: National Reform Union.

Paré, Ambroise. 1575. La Troisième Livre des Playes Faictes par Hacquebutes, Batons à Feu, Fleches, Dardes et des Accidens des Icelles, *Les Oeuvres de M. Ambroise Paré,* 368-93. Paris: Chez Gabriel Buon.

Parish, Samuel. 2008. *Central Florida's Most Notorious Gangsters: Alva Hunt and Hugh Gant.* Charleston, South Carolina: The History Press.

Pasley, Fred D. 1930. *Al Capone: The Biography of a Self-Made Man.* Salem, New Hampshire: Ayer Publishing Co.

Patrick, Urey. 1989. *Handgun Wounding Factors and Effectiveness.* U.S. Department of Justice, FBI Tactical Training Unit.[online]

Petzal, David E. 1993. Bullet-In, *Field and Stream* February: 65-7.

Preuss, J. and B. Madea. 2009. Gerhart Panning (1900-1944): A German Forensic Pathologist and His Involvement in Nazi Crimes during the Second World War, *Journal of Forensic Pathology* March 30: 14-7.

Marohomsalic, Nasser A. 2001. *Aristocrats of the Malay Race: A History of the Bangsa Moro in the Philippines.* Marawi City.

von Moltke, Helmuth James. 1988. *Briefe an Freya.* Munich: C.H. Beck.

Morris, Charles. 1873. The Domain of the Savage, *To-day: the illustrated magazine*

Ogston, Alex. 1899a. Continental Criticism of English Rifle Bullets, *British Medical Journal* 1, 1995: 752-57.

Ogston, Alex. 1899b. The Peace Conference and the Dum-Dum Bullet, *British Medical Journal* 1: 278-81.

Packard, John Hooker. 1882. *A System of Surgery, Theoretical and Practical.* 3 v. Philadelphia: Henry Lea and Son.

Panning, Gerhart. 1942. Wirkungsform und Nachweis der sowetischen Infanteriespregmunition, *Der Deutsche Militarartzt* 7: 20-30.

Parks, W. Hays. 1991. Memorandum of Law-Sniper Use of Open Tip Ammunition (12 October 1990), *The Army Lawyer* (DA Pam 27-50-218): 86-89.

Paré, Ambroise. 1968. *The Apologie and Treatise of Ambroise Paré,* edited by Geoffrey Keynes. New York: Dover.

Pegler, Martin. 2011. *Out of Nowhere: A History of the Military Sniper, From the Sharpshooter to Afghanistan.* Osprey Publications.

Peheim, Christian. 2007. Was Abu-Jamal's Weapon the Murder Weapon? *The Anti-MOVE/Mumia Blog* October 14 [online]

Pershing, John J. 1913. *The Annual Report of the Governor of the Moro Province.* Washington: Government Printing Office.

Pinizzotto, Antonio, Harry A. Kern and Edward F. Davis. 2004. *One-Shot Drops: Surviving the Myth.* FBI Law Enforcement Bulletin, October. [online]

Pogue, Forrest. 2001. *Pogue's War: Diaries of a WWII Combat Historian.* Lexington: University of Kentucky Press.

Pollard, Hugh Bertie Campbell. 1920. The Irish Gunman, *The Weekly News,* October 28 photo on website

Primley, Jacob. 2003. Columbus Raid, 213 IN *Mexico and the U.S.* Tarrytown: Marshall Cavendish.

Puckle, James. 1718. *England's Path to Wealth and Honour in a Dialogue between an English-Man and a Dutch-Man.* London: Edward Symon.

Rand, Robert. 1985. *Papa Jack: Jack Johnson and the Era of White Hopes.* New York: Simon and Schuster.

Reiss, Rodolfe Archibald. 1916. *Report Upon the Atrocities Committed by the Austro-Hungarian Army During the Austrian Invasion of Serbia,* translated by F.S. Copeland. London: Simkin, Marshall, Hamilton, Kent and Co.

Retana, Wenceslao. 1921. *Diccionario de Filipinismos.* New York and Paris.

Richards, Frank. 1936. *Old Soldier Sahib.* London: Faber and Faber. Reprinted 2003, Naval and Military Press.

Riley, Glenda. 1994. *The Life and Legend of Annie Oakley.* Norman: University of Oklahoma Press.

Rogal, William W. 2010. *Guadalcanal, Tarawa and Beyond: A Mud Marine's Memoir of the Pacific Islands War.* Jefferson, North Carolina: McFarland.

Rosenberg, Matthew. 2013. Insider Attacks in Afghanistan Shape the Late Stages of the War, *The New York Times* January 4

Russell, M.A, et al. 1995. Safety in Bullet Recovery Procedures: A Study of the Black Talon Bullet, *American Journal of Forensic Medical Pathology* 16(2): 120-3.

Salle, G.-F.-S. 1899. *Balles Humanitaires et Leurs Blessures: Conference Régimentaire Faite aux Officiers en 1897 et 1898.* Paris: Henri-Charles Lavauzelle.

Savage Landor, Arthur Henry. 1904. *The Gems of the East.* 2 v. London: Macmillan.

Sawyer, Ralph D. 2004. *Fire and Water: The Art of Incendiary and Acquatic Warfare in China.* Boulder: Westview.

Sayeg Helú, Jorge. 1996. *Paginas de la Revolucion Mexicana* I. México: Instituto Politécnico Nacional.

Schachner, August. 1900. The Surgical Aspects of the Modern Small-Bore Projectile, *Annals of Surgery* 31,1: 75-86.

Schleicher, Ron. 2006. *Psychological Warfare in the Intifada: Israeli and Palestinian Media Politics and Military Strategies.* Eastbourne: Sussex Academic Press.

Scott, Douglas, et al. 1989. *Archaeological Perspectives on the Battle of the Little Bighorn.* Norman: University of Oklahoma Press.

Slowik, Max. 2012. Hornady Adding Tons of New Products in Many New Calibers for 2013, *Guns.com* November 7 [online]

Smith, Rolling. 1897. Small-Bore Rifles for Big Game, *The Sportsman's Magazine* 1: 346-47.

Smothers, Ronald. 1993. Manufacturer Withdraws Controversial Ammunition, *The New York Times,* November 23.

Sonasundaram, Daya. 2010. Collective Trauma in the Vanni: A Qualitative Inquiry into the Mental Health of the Internally Displaced due to the Civil War in Sri Lanka, *International Journal of Mental Health Systems* 4: 22-35.

Spear, Raymond. 1906. *Report on the Russian Medical and Sanitary Features of the Russo-Japanese War.* Washington: Government Printing Office.

Spiers, Edward. 1973. The Use of the Dum-Dum Bullet in Colonial Warfare, *Journal of Colonial and Imperial History* 4,1: 3-14.

Spores, John C. 1988. *Running Amok: An Historical Study.* Athens, Ohio: Ohio University Center for International Studies.

Stangardt, K. and W. Kirschener. 1915. English Bullet Wounds: Some Remarks on the Action of the Regular Infantry Bullet and the Dumdum Bullet, *Journal of the Royal Army Medical Corps,* June: 601.

Starkey, Armstrong. 1998. *European and Native American Warfare,* 1675-1815. Abingdon: Taylor and Francis.

Steedly, Mary Margaret. 2000. Modernity and the Memory Artist: The Work of Imagination in Highland Sumatra, 1947-1995, *Comparative Studies in Society and History* 42, 4: 811-46.

Stevens, Boyd. 1994. The 101 California Street Shooting: The Black Talon Bullet, *The 1994 IWBA Conference*

Stiles, Daniel. 2011. *Elephant Meat Trade in Central Africa: Democratic Republic of Congo Case Study.* Gland, Switzerland: LUCN.

Stewart, Richard Winship. 2005. *American Military History, v. 1: The U.S. Army and the Forging of a Nation, 1775-1917.* Washington: Government Printing Office.

Swift, B. and Rutty, G.N. The Explosive Bullet, *Journal of Clinical Pathology* 57, 1: 108.

Takahashi, Sakuyé. 1908. *International Law Applied to the Russo-Japanese War.* New York: The Banks Law Publishing Company.

Taubman, Philip. 1981. Explosive Bullet Struck Reagan, FBI Discovers, *The New York Times*, March 3.

Taylor, John. 1948. *African Rifles and Cartridges.* Georgetown, South Carolina: Small Arms Technical Publishing Company (reprint of original edition published by Thomas G. Samworth).

Taracena, Alfonso. 1987. *Historia Extraoficial de la Revolución Mexicana.* Mexico: Editorial JUS.

Taylor, Chris. 2012. *The Black Carib Wars: Freedom, Survival and the Making of the Garifuna.* Oxford: Signal Books.

Thom, Deborah. 1995. Revolution in the Workplace? Women's Work in Munitions Factories and Technological Change, 1914-1918, 101-124 IN *Women Workers and Technological Change in Europe in the Nineteenth and Twentieth Centuries.* London: Taylor and Francis.

Thompson, C.M. 1898. Action of the Lee-Metford at Short Ranges, *Gailland's Medical Journal* 69:13-14.

Thompson, Roger. 1994. Britain, Germany, Australia and New Zealand in Polynesia, 71-92 IN *Tides of History: The Pacific Islands in the Twentieth Century,* edited by K.R. Howe and Robert C. Kiste. Honolulu: University of Hawai'i Press.

Tjader, Richard. 1910. *The Big Game of Africa.* New York: D. Appleton and Co.

Tomasky, Michael. 2012. Still Trying to Kill Grandma, Now With Hollow Point Bullets, *The Daily Beast* August 28 [online]

U.S. Department of the Army. Ordnance Department. 1897. *Report of the Chief of Ordnance to the Secretary of War.* Washington: Government Printing Office.

U.S. Department of the Army. Office of the Surgeon General. 1900. *The Use of the Roentgen Ray by the Medical Department of the United States Army in the War with Spain (1898).* Washington: Government Printing Office.

U.S. Congress. House of Representatives. 1903. *Annual Reports of the War Department for the Fiscal Year 1903. V. III: Reports of Department and Division Commanders.* Washington: Government Printing Office.

U.S. Congress. Senate. 1913. *Revolutions in Mexico: Hearing before a Subcommittee of the Committee on Foreign Relations.* Washington: Government Printing Office.

U.S. Congress. Senate. 1920. *Investigation of Mexican Affairs: Preliminary Report and Hearings.* v. 1. Washington: Government Printing Office.

U.S. War Department. 1905. *Annual Report of the War Department for the Fiscal Year Ended June 30, 1905, V. 12, pt. 3. Report of the Philippine Commission.* Washington: Government Printing Office.

U.S. War Department. Military Intelligence Service. 1943. *Tactical and Technical Trends,* Issue 40. Washington: Government Printing Office.

U.S. War Department. Office of the Chief of Staff. 1906. *Reports of Military Observers Attached to the Armies in Manchuria During the Russo-Japanese War (September 1, 1906), Part I.* Washington: Government Printing Office.

U.S. War Department, Office of the Surgeon-General. 1916. *Index-Catalogue of the Library of the Surgeon General's Office, United States Army.* Washington: Government Printing Office.

Uring, Nathaniel. 1726. *A History of the Voyages and Travels of Captain Nathaniel Uring.* London: W. Wilkins.

Ütterode, Ludwig. 1875. *Zur Geschichte der Heilkunde.* Berlin: Carl Henmann's Verlag.

de Varigny, Charles. 18??. *Nouvelle Geographie Moderne des Cinq Parties du Monde. t. V: Afrique-Océanie.* Paris: S.E. Girard et A. Boitte.

da Vigo, Giovanni. 1531. *Opera Domini Joannis da Vigo in Chyrurgia.* Lyon.

Vidmer, George. 1905. The Service Pistol and its Caliber. *Journal of the U.S. Cavalry Association* 16,58: 181-88.

Wei, Deborah and Rachael Kamel. 1998. *Resistance in Paradise.* Philadelphia: American Friends Service Committee.

Wey, Hamilton D. 1891. Criminal Anthropology, 274-91 IN *Proceedings of the Annual Congress of the National Prison Association of the United States, September 25-30, 1890.* Pittsburgh: Shaw Brothers.

Wienholt, Arnold. 1922. *The Story of a Lion Hunt.* London: Andrew Melrose, Ltd.

Wightman, Clare. 1995. *Women, Work and Engineering Industries, 1900-1950.* London: Longman.

Willibanks, James H. 2004. *Machine Guns: An Illustrated History of their Impact.* Santa Barbara, California: ABC-CLIO, Inc.

Wylde, Augustus Blandy. 1901. *Modern Abyssinia.* London: Methuen and Company.

Zhou, Jiahua. 1986. Gunpowder and Firearms, 184-91 IN *Ancient China's Technology and Science,* Compiled by the Institute of the History of Natural Sciences, Chinese Academy of Sciences. Beijing: Foreign Languages Press.

Zimmermann, Leo and Ilza Veith, eds. 1993. *Great Ideas in the History of Surgery.*

San Francisco: Norman.

Yeğen, Mesut. 2011. The Kurdish Question in Turkey: Denial to Recognition, 67-84 IN *Nationalisms and Politics in Turkey: Political Islam, Kemalism and the Kurdish Question.* Abingdon: Routledge.

Yule, Henry and A.C. Burnell. 1968 (1903). *Hobson-Jobson: A Glossary of Anglo-Indian Colloquial Words and Phrases.* New York: Humanities Press.

Index

www.ingramcontent.com/pod-product-compliance
Lightning Source LLC
Chambersburg PA
CBHW051458170526
45166CB00001B/299